LIQUID RULES

ALSO BY MARK MIODOWNIK

Stuff Matters

LIQUID RULES

The DELIGHTFUL and DANGEROUS SUBSTANCES that FLOW THROUGH OUR LIVES

MARK MIODOWNIK

HOUGHTON MIFFLIN HARCOURT

BOSTON   NEW YORK   2019

2/19

First U.S. edition

For information about permission to reproduce selections from this book,
write to trade.permissions@hmhco.com or to Permissions, Houghton Mifflin Harcourt
Publishing Company, 3 Park Avenue, 19th Floor, New York, New York 10016.

hmhco.com

First published with the title *Liquid* in Great Britain in 2018 by Penguin Books Ltd.

Library of Congress Cataloging-in-Publication Data
Names: Miodownik, Mark, author.
Title: Liquid rules : the delightful and dangerous substances that flow
through our lives / Mark Miodownik.
Description: Boston : Houghton Mifflin Harcourt, 2019. | Originally published
in Great Britain by Penguin Books, 2018. | Includes bibliographical
references and index.
Identifiers: LCCN 2018024866| ISBN 9780544850194 (hardcover) |
ISBN 9780544850200 (ebook)
Subjects: LCSH: Liquids — Popular works. | Matter — Properties — Popular works.
Classification: LCC QC145.2 .M565 2019 | DDC 530.4/2 — dc23
LC record available at https://lccn.loc.gov/2018024866

Printed in the United States of America
DOC 10 9 8 7 6 5 4 3 2 1

In loving memory of my Mum & Dad

CONTENTS

INTRODUCTION

PEANUT BUTTER, HONEY, PESTO sauce, toothpaste, and, most painfully, a bottle of single-malt whiskey—these are just some of the liquids I've had confiscated at airport security. I inevitably lose the plot in such situations. I say things like "I want to see your supervisor" or "Peanut butter is not a liquid," even though I know it is. Peanut butter flows and assumes the shape of its container—that is what liquids do—and so peanut butter is one. Even so, it just infuriates me that in a world full of "smart" technology, airport security still can't tell the difference between a liquid spread and a liquid explosive.

Although bringing more than 3.4 ounces of liquid through security has been banned since 2006, our detection technology has not improved much in that time. X-ray machines can see through your luggage to detect objects. In particular they alert security to suspicious shapes: distinguishing guns from hairdryers, and knives from pens. But liquids don't have a shape. They just take the form of whatever is containing them. Airport scanning technology can also detect density and a range of chemical elements. But here again they run into problems. The molecular makeup of the explosive nitroglycerin, for instance, is similar to peanut butter's. They're both made from carbon, hydrogen, nitrogen, and oxygen—and yet one is a liquid explosive while the other is just, well, delicious. There are an enormous number of dangerous toxins, poisons, bleaches, and pathogens that are incredibly difficult to distinguish from more innocent liquids in a quick and reliable manner. And it is this line of argument, which I've heard from many security guards (and their supervisors), that usually persuades me to agree — begrudgingly — that my peanut butter, or one of the other liquids I seem to regularly forget to take out of my carry-on luggage, is a significant risk.

Liquids are the alter ego of dependable solid stuff. Whereas solid materials are humanity's faithful friend, taking on the permanent shapes of clothes, shoes, phones, cars, and indeed airports, liquids are fluid; they will take on any shape, but only while contained. When they're not contained, they are always on the move, seeping, corroding, dripping, and escaping our control. If you put a solid material somewhere, it stays there — barring forcible removal — often doing something very useful, like holding up a building or supplying electricity for a whole community. Liquids, on the other hand, are anarchic: they have a knack for destroying things. In a bathroom, for instance, it is a constant battle to keep water from seeping into cracks and pooling under the floor, where it gets up to no good, rotting and undermining the wooden joists; on a smooth tile floor, water is the perfect slip hazard, causing an enormous number of injuries; and when it gathers in the corners of the bathroom, it can harbor black slimy fungus and bacteria, which threaten to infiltrate our bodies and make us sick. And yet, despite all this treachery, we love the stuff; we love bathing in water, and showering in it, drenching the whole body. And what bathroom is complete without a cornucopia of bottled liquid soaps, shampoos, and conditioners, jars of cream, and tubes of toothpaste? We delight in these miraculous liquids and yet we worry about them too: Are they bad for us? Do they cause cancer? Do they ruin the environment? With liquids, delight and suspicion go hand in hand. They are duplicitous by nature, neither a gas nor a solid but something in between, something inscrutable and mysterious.

Take mercury, for instance, which has delighted and poisoned humanity for thousands of years. As a child I used to play with liquid mercury, flicking it around on tabletops, fascinated by its otherworldliness, until I was made aware of its toxicity. But in many ancient cultures mercury was thought to prolong life, heal fractures, and maintain good health. It's not clear why it was held in such regard — perhaps because it is special in being the only pure metal that is liquid at room temperature. The first emperor of China, Qin Shi Huang, took mercury pills for his health but died at thirty-nine, probably as a result. Even so, he was buried in a tomb full of rivers of flowing mercury. The ancient Greeks used mercury in ointments, and alchemists believed that a combination of mercury and another elemental substance, sulfur, formed the basis of all metals — a perfect balance between mercury and sulfur creating gold. This was the origin of the erroneous belief that different metals could be transmuted into gold if mixed in the right proportions. While that proved to be the stuff of legend, gold does dissolve in mercury. If you heat up that liquid once it has absorbed the metal, it will evaporate, leaving behind a solid lump of gold. For most ancient people, that process was indistinguishable from magic.

Mercury isn't the only liquid that can consume another substance and contain it within itself. Add salt to water and it will soon disappear — the salt is somewhere, but where has it gone? Yet if you do the same with oil, the salt just sits there. Why? Liquid mercury may absorb solid gold, but it rejects water. Why is that? Water absorbs gases, including oxygen, and if it didn't, we would live in a very different world — it is the oxygen dissolved in water that allows fish to breathe. And although water can't carry enough oxygen for humans to breathe, other liquids can. There's a type of oil — perfluorocarbon liquid — that is very unreactive chemically and electrically. It is so inert that you can put your cellphone in a beaker of perfluorocarbon liquid and it will continue to operate normally. Perfluorocarbon liquid can also absorb oxygen in such high concentrations that it is breathable by humans. This sort of liquid breathing — breathing liquid instead of air — has many possible uses, the most important of which is to treat premature babies suffering from respiratory-distress syndrome.

Still, it is liquid water that has the ultimate life-giving property. This is because it can dissolve not just oxygen, but many other chemicals too, including carbon-based molecules, and so provides the aqueous environment required for the emergence of life — for new organisms to spontaneously come into existence. Or at least that is the theory. And it is why, when scientists look for life on other planets, they look for liquid water. But liquid water is rare in the universe. It's possible that Europa, one of the moons of Jupiter, might have oceans of liquid water below its icy crust. There might also be liquid water on Enceladus, one of the moons of Saturn. But Earth is the only body in the solar system where a lot of water is readily available on the surface.

A peculiar set of circumstances on our planet has made possible the surface temperatures and pressures that can sustain liquid water. In particular, if it weren't for Earth's molten-metal liquid core, which creates the magnetic field that protects us from the solar wind, all our water would most likely have disappeared billions of years ago. In short, on our planet, liquid begat liquid, and that led to life.

But liquids are destructive too. Foams feel soft because they are easily compressed; if you jump on to a foam mattress, you'll feel it give beneath you. Liquids don't do this; instead they flow — with one molecule moving into the space freed up by another molecule. You see this in a river, or when you turn on a tap, or if you use a spoon to stir your coffee. When you jump off a diving board and hit a body of water, the water has to flow away from you. But the flowing takes time, and if your speed of impact is too great, the water won't be able to flow away fast enough, and so it pushes back at you. It's that force that stings your skin as you belly-flop into a pool, and makes falling into water from a great height like landing on concrete. The incompressibility of water is also why waves can exert such deadly power, and in the case of tsunamis, why they can demolish buildings and cities, tossing cars around as if they were driftwood. For instance, the Indian Ocean earthquake in

2004 triggered a series of tsunamis, killing 230,000 people across fourteen countries. It was the eighth-worst natural disaster ever recorded.

Another dangerous property of liquids is their ability to explode. When I began my Ph.D. at Oxford University, I had to prepare small specimens for the electron microscope. This involved cooling a liquid called an electropolishing solution to a temperature of –4°F. The liquid was a mixture of butoxyethanol, acetic acid, and perchloric acid. Another Ph.D. student in the lab, Andy Godfrey, showed me how to do this, and I thought I'd gotten the hang of it. But after a few months Andy noticed that I often let the temperature of the solution rise while electropolishing. "I wouldn't do that," he said, raising his eyebrows as he peered over my shoulder one day. When I inquired why, he pointed me toward the lab manual of chemical hazards:

> Perchloric acid is a corrosive acid and destructive to human tissue. Perchloric acid can be a health hazard if inhaled, ingested or splashed on skin or eyes. Once heated above room temperature or used at concentrations above 72 per cent (any temperature), perchloric acid becomes a strong oxidizing acid. Organic materials are especially susceptible to spontaneous combustion if mixed or contacted with perchloric acid. Perchloric acid vapours may form shock-sensitive perchlorates in ventilation system ductwork.

In other words, it can explode.

Upon inspection of the lab I found many similarly transparent colorless liquids, most of which were indistinguishable from one another. We used hydrofluoric acid, for instance, which, apart from being an acid that can eat its way through concrete, metals, and flesh, is also a contact poison that interferes with nerve function. This has an insidious consequence — namely, that you can't feel the acid as it's burning you. Accidental exposure can easily go unnoticed as the acid continues to eat its way through your skin.

Alcohol too fits into the category of poison. It may be poisonous only in high doses, but it has killed many more people than hydrofluoric acid has. Yet alcohol plays an enormous role in societies and cultures around the globe, having historically been used as an antiseptic, an antitussive, an antidote, a tranquilizer, and a fuel. Alcohol's main attraction is that it depresses the nervous system — it's a psychoactive drug. Many people can't function without their daily glass of wine, and most social functions revolve around places where alcohol is served. We (rightly) may not trust these liquids, but we love them anyway.

We feel alcohol's physiological effects as it's absorbed into the bloodstream. The thumping of our heartbeat is a constant reminder of blood's role in the body and its need to constantly circulate: we run thanks to the power of a pump, and when the pumping stops, we die. Of all the liquids in the world, surely blood is one of the most vital. Fortunately, the heart can now be replaced, bypassed, and plumbed in and out of the body. Blood itself can be added and removed, stored, shared, frozen, and revived. And indeed, without our blood banks, every year millions of people undergoing surgery, injured in war zones, or involved in traffic accidents would die.

But blood can be contaminated with infections such as HIV and hepatitis, and so it can harm as well as heal. Thus, we must take into consideration the duplicitous nature of blood, as of all liquids. The important question is not whether a particular liquid can be trusted or not, is good or bad, is healthy or poisonous, is delicious or disgusting, but rather whether we understand it enough to harness it.

There is no better way to illustrate the power and delight we gain from controlling liquids than by taking a look at those involved in the flight of an airplane and the experience of the passengers onboard. And so that is what this book is about, a transatlantic flight, and all the strange and wonderful liquids that play a part in it. I took this flight because I had not blown myself up while completing my Ph.D., but had instead continued to do materials science research and eventually become director of the Institute of Making, at University College London. There, part of our research involves understanding how liquids can masquerade as solids. For instance, the tar from which roads are made is, like peanut butter, a liquid, even though it gives the impression of being a solid. Our research has led to invitations to fly to conferences all around the world, and this book is an account of one such trip, from London to San Francisco.

The flight is described through the language of molecules, heartbeats, and ocean waves. My aim is to unlock the mysterious properties of liquids and how we have come to rely on them. The flight takes us over the volcanoes of Iceland, the frozen expanse of Greenland, the lakes dotted around Hudson Bay, and then south to the coastline of the Pacific Ocean. This canvas is big enough to accommodate a discussion of liquids from the scale of oceans down to droplets in the clouds, along with the curious liquid crystals in the onboard entertainment system, the beverages served by the flight attendants, and, of course, the aviation fuel that keeps a plane in the stratosphere.

In each chapter I consider a different part of the flight and the qualities of liquids that made it possible: their ability to combust, to dissolve, or to be brewed, to name a few. I show how wicking, droplet formation, viscosity, solubility, pressure, surface tension, and many other strange properties of liquids can allow us to fly around the globe. And in doing so I reveal why liquids flow up a tree but down a hill, why oil is sticky, how waves can travel so far, why things dry, how liquids can be crystals, how not to poison yourself making hooch, and, most important perhaps, how to make the perfect cup of tea. So please, come fly with me — I can promise you a strange and marvelous trip.

LIQUID RULES

1 / EXPLOSIVE

AS SOON AS THE aircraft doors closed and we pushed back from the gate at Heathrow Airport, a voice announced the beginning of the preflight safety briefing.

"Good afternoon, ladies and gentlemen, and welcome to this British Airways flight to San Francisco. Before our departure, may we have your attention while the cabin crew point out the safety features aboard this airplane."

I always find this a disconcerting way to start a flight. I am convinced that it's a fake: that the safety briefing isn't really about safety at all. For a start, it fails to mention the tens of thousands of gallons of aviation fuel onboard. It is the enormous amount of energy contained in this liquid that allows us to fly at all; its fiery nature is what powers the jet engines so that they're capable of taking, in our case, four hundred passengers in a 275-ton aircraft from a standing start on the runway to a cruising speed of five hundred miles an hour, and to a height of forty thousand feet, in a matter of minutes. The sheer awesome power of this liquid fuels our wildest dreams. It allows us to soar above the clouds and travel anywhere in the world in a matter of hours. It's the same stuff that took the first astronaut, Yuri Gagarin, into space in his rocket, and that fuels the latest generation of SpaceX rockets, which fire satellites into the atmosphere. It is called kerosene.

Kerosene is a transparent, colorless fluid that, confusingly, looks exactly like water. So where is all that hidden energy stored, all that secret power? Why doesn't the storage of all that raw energy

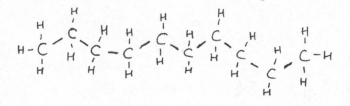

A sketch of the structure of a hydrocarbon molecule in kerosene.

inside the liquid make it appear, well, more syrupy and dangerous? And why is it not mentioned in the preflight safety briefing?

If you were to zoom in and have a look at kerosene on the atomic scale, you would see that its structure is like spaghetti. The backbone of each strand is made of carbon atoms, with each one bonded to the next. Every carbon is attached to two hydrogen atoms, except at the ends of the molecule, which have three hydrogen atoms. At this scale you can easily tell the difference between kerosene and water. In water there isn't a spaghetti structure but rather a chaotic jumble of small V-shaped molecules (one oxygen atom attached to two hydrogen atoms, H_2O). No, at this scale kerosene more closely resembles olive oil, which is also comprised of carbon-based molecules all jumbled up together. But where the strands in kerosene are more like spaghetti, in olive oil they're branched and twirled.

Because the molecules in olive oil have a more complex shape than the ones in kerosene, it's harder for them to wiggle past each other, and so the liquid flows less easily — in other words, olive oil is more viscous than kerosene. They're both oils, and on the atomic level they look relatively similar, but because of their structural differences, olive oil is gloopy while kerosene pours more like water. This difference doesn't just determine how viscous these oils are but also how flammable.

The Persian physician and alchemist Rhazes wrote about his discovery of kerosene in his ninth-century *Book of Secrets*. Rhazes became interested in the naturally occurring springs of the region, which oozed not water but a thick black sulfurous liquid. At the time, this tarlike material was extracted and used on roads, essentially as an ancient form of asphalt. Rhazes developed special chemical procedures, which we now call distillation, to analyze the black oil. He heated it up and collected the gases that were expelled from it. He then cooled these gases down again, whereupon they were transformed back into liquid. The first liquids he extracted were yellow and oily, but through repeated distillation they became a clear, transparent, and free-flowing substance — Rhazes had discovered kerosene.

At the time Rhazes couldn't know the full extent of what this liquid would ultimately contribute to the world, but he did know it was flammable and that it produced a smokeless flame. While this may seem like a trivial discovery now, creating indoor light was a major problem for every ancient civilization. Oil lamps were the most sophisticated light-producing technology of the day, but up until then, burning oil often produced as much soot as it did light. Smokeless oil lamps would be a revolutionary innovation, so much so that their importance is immortalized in the story of Aladdin, from *The Book of One Thousand and One Nights*. In the story, Aladdin finds an oil lamp, a magic lantern. When he rubs it, he releases a powerful genie. Genies occur frequently in myths of the time and are said to be supernatural creatures made from a smokeless fire; this particular genie is bound to do the bidding of the person who owns the lamp — an immense power. The significance of the new liquid and its ability to create a smokeless flame could not have been lost on the alchemist Rhazes. So why didn't the Persians start using this new spirit? The answer comes, in part, from the importance that olive trees had in their economy and culture.

In the ninth century, olive oil was the fuel of choice for oil lamps in Persia. Olive trees thrived in the region; drought-tolerant, they yielded olives, which could be pressed into oil. It took about twenty olives to create a teaspoon of olive oil, which provided one hour of light with a typical oil lantern. So, if an average household needed five hours of light per evening, it would go through a hundred olives a day, or approximately thirty-six thousand olives a year, just to light one lamp. To produce enough oil to illuminate their empire, the Persians needed an abundance of both land and time, because olive trees mature slowly; generally, they don't produce fruit for their first twenty years. The Persians also needed to protect their land from anyone who might want to take this valuable resource, which meant they needed organized towns, and this required still more olives, so everyone could have both cooking oil and light. To support an army the people needed to pay taxes, and in Persia paying taxes often meant giving the government a percentage of the olive oil you produced. So, you can see, olive oil was central to Persian society and culture, as it was to all Middle Eastern civilizations, until they found an alternative source of energy and tax revenue. Rhazes's experiments proved that it was right under their feet, but there it would stay for another thousand years.

In the meantime, oil lamps evolved. The design of a ninth-

A replica of an ancient oil lamp used at the time of Rhazes.

century oil lamp looks simple, but it is remarkably sophisticated. Think about a bowl of olive oil. If you simply try to light it, you'll find it's quite difficult. It's hard because olive oil has a very high flashpoint—the temperature at which a flammable liquid will spontaneously react with the oxygen in the air and burst into flames. For olive oil this is 600°F. That's why cooking with olive oil is so safe. If you spill it in your kitchen, it's not going to ignite. Also, to fry most foods you need to get to a temperature of only around 400°F, which is still about two hundred degrees below olive oil's flashpoint, so it's easy to cook without the oil burning.

But at 600°F, your pot of olive oil will burst into flames and, in doing so, will create a lot of light. Not only is this incredibly dangerous, but the flames will be short-lived; they'll consume all the fuel very quickly. Surely, you must be thinking, there's a better way to burn olive oil for light. And as it turns out, there is. You don't have to heat up the full pot of oil. If you take a piece of string, submerge it in the oil, leaving the top poking above the surface, and then light it, a bright flame is created at the tip of the string. It is not the string that creates the flame; it is the oil emerging from it. This is ingenious, but it gets better. If you continue to let it burn, the flame doesn't travel down into the oil—instead, the oil climbs up the string, igniting only when it gets to the top. This system can maintain the flame for hours; indeed, for as long as there is oil in the bowl. It's a process called wicking and seems miraculous —the oil is able to defy gravity and move autonomously—but it's a basic principle of liquids and it's possible because they possess something called surface tension.

What gives liquids their ability to flow is their structure—they are an intermediate state between the chaos of gas and the static prison (for molecules) of solids. In gases, molecules have enough heat energy to break away from one another and move autonomously. This makes gases dynamic—they expand to fill the available space—but they have almost no structure. In solids the force of attraction between the atoms and molecules is much greater than the heat energy they possess, causing them to bond together. Thus, solids have a lot of structure but their atoms have little autonomy—when you pick up a bowl, all the atoms of that bowl come together as one object. Liquids are an intermediate state between the two. The atoms have enough heat energy to break some of the bonds with their neighbors but not enough to break all of them and become a gas. So they are stuck in the liquid but able to move around within it. This is what a liquid is—a form of matter in which molecules swim around, making and breaking connections with one another.

Molecules at the surface of a liquid and those inside it experience different environments. Molecules at the surface are not completely surrounded by other molecules, so they undergo less bonding on average than those in the middle of the liquid. This imbalance of forces between the surface and the interior of the liquid creates a force of tension—called surface tension. The force is tiny, but it's still big enough to oppose the force of gravity on small things: this is why some insects are able to walk on the surface of ponds.

A pond skater insect walking on water.

Look carefully at the pond skater insect as it "walks" on water, and you'll see that its legs are repelled by the water — this happens because the surface tension between the water and the insect's legs generates a repulsive force that acts against gravity. Some liquid-solid interactions do the opposite and create a molecular force of attraction. This is true of water and glass. Take a glass of water and you will see that the edges of the water are pulled up where they meet the glass. We call this the meniscus, and it too is a surface-tension effect.

Plants have mastered this same trick. They pull water up against the force of gravity, from the ground into their bodies, using a system of tiny tubes that run through their roots, stems, and leaves. As the tubes become microscopic, so the ratio of the tube's inner surface area to the volume of liquid increases, and so the effect gets bigger. Hence, manufacturers sell "microfiber" cloths for window cleaning, which have microchannels similar to a plant's. They suck up water, allowing the cloth to clean more efficiently. Paper towels mop up liquid spills using the same mechanism. These are all examples of "wicking" — the same surface-tension effect that allows oil to climb up a string — or, more precisely, a wick.

Without wicking, candles wouldn't work. When you light the wick on a candle, the heat melts the wax and creates a pool of molten wax. This liquid wax travels up the wick, through microchannels, to the flame, thus feeding it with a new supply of liquid wax to burn. If you choose the right material for the wick, the flame will burn hot enough to maintain a pool of liquid wax and ensure a constant flow of fuel. This familiar self-regulating system requires so little input from us, we fail to regard a candle as a complex piece of technology, but that is exactly what it is.

For thousands of years, all around the globe, wicking provided the primary mechanism for indoor lighting, whether in candles or oil lamps. For those who lacked these technologies, at night the world descended into a dark gloom. As you might expect, oil lamps were popular in places where oils were plentiful, while candles were used where wax or animal fat was more readily available.

Nevertheless, as ingenious as they were, candles and oil lamps had their drawbacks: obviously, the fire risk, but also the production of soot, the low brightness of the flame, the smell, and the cost. These shortcomings meant that there were always those searching for better and cheaper and safer ways of providing indoor light. Rhazes's discovery of kerosene in the ninth century would have been the solution, had anyone realized it.

Onboard the aircraft the preflight safety briefing was in full swing, and now the flight attendants too were ignoring the importance of kerosene. There had not been the least mention of it so far, even though this revolutionary stuff was, at that very moment, being sprayed into the jet engines under the plane's wings to power our aircraft as it taxied to the runway. Instead, the attendants were talking about what to do in the event of "loss of cabin pressure." As an Englishman I appreciate the understated nature of this phrase. It sounds like no big deal, but what it means is that while cruising at high altitude, if the cabin suddenly developed a hole or a crack, all the air would be sucked out of the aircraft, along with anyone not strapped into their seat. There wouldn't be enough oxygen for people to breathe normally; hence the oxygen masks that are designed to drop down from the ceiling. The aircraft would immediately begin a steep descent to reach lower altitudes, where there is more oxygen. Anyone left alive at that point would indeed be safe.

Lack of oxygen was a problem for ancient oil lamps too. The design didn't allow enough oxygen to get to the fuel to burn it fully, which is why the flame gave off relatively low light. This was a problem right up until the eighteenth century, when a Swiss scientist named Ami Argand invented a new type of oil lamp, with a sleeve-shaped wick protected by a transparent glass shield. It was designed so that air could pass through the middle of the flame, radically improving the amount of oxygen delivered and thus the efficiency and brightness of oil lamps, making one equivalent to six or seven candles. This innovation led to many more, and eventually it became clear that olive oil and other vegetable oils were not ideal fuels. To get brighter light, you need higher temperatures, and for that you need faster wicking, and the speed of the wicking is determined by the surface tension and the viscosity of the liquid. Trying to find oils that were cheap but also had a low viscosity led to more experimentation, and, sadly, the deaths of a lot of whales.

Capturing a Sperm Whale, by John William Hill (1835).

Whale oil is produced by boiling strips of whale blubber. The oil the blubber releases is a clear honey color. It's not great for cooking or eating because of its strong fishy taste, but with a flashpoint of 446°F and low viscosity, it is very good for oil lamps.

The use of whale oil in Argand lamps skyrocketed in the late eighteenth century, especially in Europe and North America. From 1770 through 1775 the whalers of Massachusetts produced forty-five thousand barrels of whale oil per year to meet the demand. Whaling became a big industry, fueled by the need for indoor lighting, and some species of whales were almost driven to extinction by that demand. It's estimated that by the nineteenth century more than a quarter of a million whales had been killed for their oil.

This could not go on, and yet the demand for indoor lighting was still on the rise. As populations grew bigger and wealthier, education became more important, the culture of reading and entertaining after dark became more prevalent, and so the demand for oils increased, as did the pressure on inventors and scientists to come up with ways to meet this need. Among them was James Young, a Scottish chemist who, in 1848, found a way to extract a liquid out of coal that had excellent properties for burning in an oil lamp. Young called his liquid paraffin oil. A Canadian inventor, Abraham Gesner, did the same thing and called his product kerosene. These discoveries might not have come to much, but as it turned out, they just preceded the beginning of the American Civil War. Whaling ships became military targets, and taxes on other lamp oils created an opportunity for this new kerosene industry to find a foothold. But it didn't really take off until inventors started playing around not with coal but with the black oil that could be found near coal mines. This crude oil, which had to be pumped out of the ground, is a black, smelly, sticky substance. But before people could use it, they had to harness distillation, an old trick first used by Rhazes — which proved to be extremely lucrative. Now the genie really was out of the lamp.

Meanwhile, onboard my plane, there was still no word of kerosene. The safety briefing had gotten to the bit about emergency exits, and the flight attendant in front of me was shooting out his arms, fingers extended to identify their locations. There were two exits behind me, and two at the front of the aircraft, and two over the wings, I was told. I wanted to add this: "And there are about 13,000 gallons of kerosene in the fuel tank below our feet, and another 13,000 gallons stored in each of the two wings of the aircraft." I must have muttered something to this effect because I attracted the attention of my neighbor, whose name I would later find out was Susan. For the first time since she had gotten on the plane, Susan looked up from her book. She caught my eye for the briefest of moments over the top of her red-rimmed glasses and then returned to her reading. Her glance must have lasted less than a second, but it spoke volumes. It said, "Relax. Flying is the safest form of long-distance travel — do you know that every day there are more than a million humans flying in the stratosphere? — the chance of anything bad happening is minuscule. No, it's smaller than minuscule. Sit back. Relax. Read a book." I know this is a lot of information for a look to convey, but believe me, hers said all of that.

An oil refinery; the tall columns are distillation vessels.

For better or worse, though, all I could think about was kerosene, and the remarkable trick those mid-nineteenth-century inventors used to transform crude oil: distillation. In order to distill oil, Rhazes had used an apparatus called an alembic, which is what, in modern times, we call distillation vessels — the towers you see sticking up out of oil refineries.

Crude oil is a mixture of differently shaped hydrocarbon molecules, some long like spaghetti, some smaller and more compact, others linked together in rings. The backbone of each molecule is made of carbon atoms, each one bonded to the next. Each carbon atom also has two hydrogen atoms attached to it, but there's a lot of variety in their shape and size: the molecules vary in size from just five carbon atoms to hundreds. There are very few hydrocarbon molecules with fewer than five carbon atoms, though, because molecules that small tend to exist as gases: they're called methane, ethane, and butane. The longer the molecule, the higher its boiling point, so the more likely it is to be a liquid at room temperature. This is true of hydrocarbon molecules made up of up to forty carbon atoms. If they get any bigger than that, they can hardly flow at all and so become a tar.

In distilling crude oil, the smaller molecules are extracted first. Hydrocarbon molecules with five to eight carbon atoms form a clear transparent liquid that is extremely flammable; it has a flashpoint of −49°F, which means that even at subzero temperatures it will ignite easily. So easily, in fact, that putting this liquid into

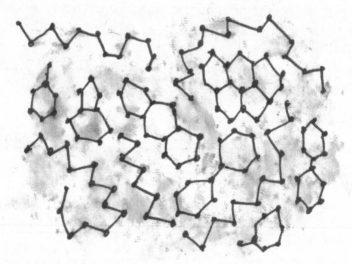

A mixture of hydrocarbon molecules contained in crude oil (only the carbon atoms are shown).

an oil lamp is quite dangerous. Thus, in the early days of the oil industry, it was discarded as a waste product. Later, when we began to better understand the virtues of this liquid, it became more appreciated, especially for the way it mixed with air and ignited, producing enough hot gas to drive a piston. It was subsequently named gasoline, and we began using it to fuel gas engines.

Larger carbon molecules, with nine to twenty-one carbon atoms, form a transparent clear liquid with a boiling point higher than gasoline's. It evaporates at a slow rate and so is less easy to ignite. But because each molecule is quite big, when it does react with oxygen, it gives off a lot of energy, in the form of hot gas. It won't ignite, however, unless it's sprayed into the air, and it can be compressed to a high density before it bursts into flames. This is the principle Rudolf Diesel discovered in 1897, eventually giving his name to the liquid forming the basis of his tremendous invention: the most successful engine of the twentieth century.

But in the early days of the oil industry, the mid-nineteenth century, diesel engines hadn't been invented yet, and there was a pressing need for a flammable substance for oil lamps. While searching for this oil, producers created a liquid that had carbon molecules with six to sixteen carbon atoms. This liquid is somewhere between gasoline and diesel. It has the virtues of diesel—it doesn't evaporate so quickly as to form explosive mixtures—but it is a fluid with a very low viscosity, similar to that of water. As a result it wicks extremely well, allowing a flame to be very bright. It was cheap and effective, and didn't rely on olive trees or whales. It was kerosene, the perfect lamp oil.

But is it safe? My mind had wandered—I had been trying to relax as per Susan's implicit instruction—but now my attention snapped back to the flight attendants again. They had gotten to the bit in the safety briefing about the life jackets. Each was now wearing one while pretending to blow a whistle. I wondered what it would feel like to survive a crash landing on the sea and be floating in water, perhaps at night, trying to blow the whistle. I also

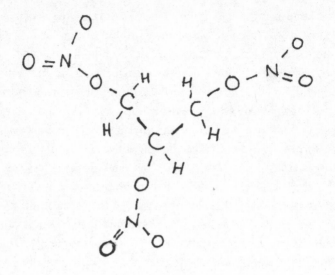

The molecular structure of nitroglycerin.

wondered what would happen to the kerosene in our fuel tanks in the event of such a crash. Could it explode?

I know one liquid that certainly could: nitroglycerin. Like kerosene, nitroglycerin is a colorless, transparent, oily liquid. It was first synthesized by the Italian chemist Ascanio Sobrero in 1847. It didn't kill him, which is a miracle, because it is a ridiculously dangerous and unstable chemical that can explode unexpectedly. Ascanio was so frightened by the potential uses of what he had discovered, he kept it quiet for a year and even then tried to deter others from making it. His student Alfred Nobel saw its potential, though; he thought it could replace gunpowder. He eventually succeeded in creating it in a form that was relatively safe to handle. Alfred transformed the liquid into a solid that wouldn't explode accidentally (although it did kill his brother Emil) and so created dynamite. This transformed the mining industry, making Alfred a fortune. Prior to dynamite, mining companies had relied on manual labor to dig their tunnels, pits, and caverns. He used his fortune—or at least a part of it—to create the most famous international awards in the world, the Nobel Prizes.

Like gasoline, diesel, and kerosene, nitroglycerin is made from carbon and hydrogen. But it comes with extras too: oxygen and nitrogen atoms. The presence of these atoms, and their positions within the molecule, make nitroglycerin unstable. If the molecule comes under pressure from contact or vibration, it can easily fall apart. When this happens, the nitrogen atoms get together to form a gas, and the oxygen atoms in the molecule react with the carbon to form carbon dioxide, another gas. They also react with the hydrogen to form steam, and whatever is left over forms still more oxygen gas. As the molecule decomposes, it creates a shock wave in the nitroglycerin, which causes the neighboring molecules to fall apart too, creating more gas and sustaining the shock wave. Ultimately, all of the nitroglycerin molecules decompose in a chain reaction that occurs at thirty times the speed of sound, transforming the liquid into a hot gas almost instantaneously. This gas has a volume a thousand times the volume of the liquid and so it expands rapidly, causing an enormous hot explosion. Much of the devastation of World War II was caused by the widespread use of nitroglycerin-based explosives.

The 3.4-ounce limit on liquids carried onto airplanes is designed to prevent someone from bringing onboard a large enough quantity of a liquid explosive, such as nitroglycerin, to destroy a plane. This amount of nitroglycerin will still explode, of course, but not with enough energy to bring the plane down. But still, it is sobering to think that kerosene contains ten times more energy per gallon than nitroglycerin, and there are tens of thousands of gallons of it in the fuel tanks of an airplane.

Kerosene is not an explosive, though—it will not spontaneously blow up. Unlike nitroglycerin, it doesn't have any oxygen and nitrogen atoms in its molecular structure. It is a stable molecule that doesn't readily decompose. You can bash it, smash it, or have a bath in it, and it won't explode. In this, it differs from its less powerful cousin, nitroglycerin. If you want to harness the power of kerosene, you have to work for it — you need to make it react with oxygen. As the kerosene and the oxygen react, they will create carbon dioxide and steam, but because the reaction is limited by its access to oxygen, the combustion can be controlled.

It is the huge power of kerosene, and our ability to burn it in a controlled manner, that makes it such an important liquid technologically. Global civilization currently burns approximately 250 million gallons of kerosene per day, mostly in jet engines and space rockets, but it is also still used for lighting and heating in many countries. In India, for instance, more than 300 million people use kerosene oil lamps to provide lighting in their homes.

Still, as much as we like to think we've got kerosene under control, there's nonetheless a sinister side to it. The horrors of September 11, 2001, are a case in point. On that day I was at home, staring in disbelief at the television. In truth I can't remember if I saw live footage of the second plane flying into one of the Twin Towers or whether what I saw was a news recap, but it stunned me. I stood looking dumbfounded at the TV, trying to comprehend the scene. The two buildings were on fire, and there were reports of other planes being flown into targets elsewhere. It seemed like things couldn't get any worse, and then they did: the first tower came down, collapsing in the type of slow motion that only giant objects can sustain. And then the second tower came down. We were ready for it this time, but it was no less numbing.

It was the fuel from the aircraft that caused the towers to collapse. It wasn't an explosion, because kerosene is stable. According to the FBI report, the kerosene reacted with oxygen from the winds blowing through the buildings' damaged floors, raising the temperature on those floors to over 1,500°F. This did not melt the steel frame of the building — steel melts at temperatures exceeding 2,500°F — but at 1,500°F, the strength of steel decreases approximately by half, and so it started to buckle. Once one floor buckled, the weight of the entire building above it collapsed onto the floor below, causing it to buckle, and so on, like a house of cards. In total, more than 2,700 people were killed in the collapse of the Twin Towers, including 343 New York firefighters. These terror attacks were a significant moment in the history of world, not just because they initiated wars and all the horrors that go with them, but because the fall of those towers was a powerful symbol of the fragility of democratic civilization. And the active ingredient of that moment of destruction was the planes' kerosene.

So you can see why I would think the flight attendants might mention it in our safety briefing. But it had just ended, and they had not said a thing about the thirty-nine thousand gallons of kerosene onboard, nor commented on its dual nature: how, on the one hand, it's a perfectly ordinary transparent oil, one so stable that you could throw a lighted match into the fuel tank and it wouldn't ignite; and yet how, on the other, when mixed with the right amount of oxygen, it becomes an oil ten times as powerful as the explosive nitroglycerin. My neighbor, Susan, seemed unfazed by this; she was still deeply engrossed in her book.

Although kerosene is not mentioned explicitly in the preflight safety briefing, it occurs to me that it is nevertheless hidden in there somehow. If you think about it, the safety briefing is the one global ritual that we all share, whatever our ethnicity, nationality, sex, or religion; we all take part in it before the kerosene is ignited and the plane takes off. The dangers that the briefing warns us of, such as landing on water, are so rare that even if you flew every day for a whole lifetime, you would be unlikely to ever experience them. So that's not really the point of it. As in all rituals, coded language, a series of actions, and special props play their part. In religious rituals these props are often candles, incense burners, and chalices; in the preflight safety ritual they are oxygen masks, life jackets, and seat belts. The message of the preflight ritual is this: you are about to do something that is extremely dangerous, but engineers have made it almost completely safe. The "almost" is emphasized by all the elaborate actions involving the previously mentioned props. The ritual draws a line between your normal life, where you are in charge of your own safety, and your current one, in which you are ceding control to a set of people and their engineering systems as they harness one of the most awesomely powerful liquids on the planet to shoot you through the atmosphere to a destination of your choosing. In other words, you need to trust them absolutely; your life is in their hands. And so this ritual, performed before every flight, is really a trust ceremony.

As the cabin crew began moving down the aisles, ostensibly checking that passengers' seat belts were correctly fitted and bags were stowed, I knew that the safety ritual was coming to a close—this was the final blessing. I nodded to the attendant solemnly. The aircraft had arrived at the runway and begun its takeoff procedure, and so the accumulated knowledge of more than a thousand years was being brought to bear to turn liquid kerosene into flight.

If you have ever blown up a balloon and then let it go, allowing it to zoom and fart its way around a room, you have a good grasp of how a jet engine works. As compressed air shoots out in one direction, the balloon is propelled in the opposite: this is Newton's third law of motion, which states that every action has an equal and opposite reaction. But storing enough compressed gas to power an aircraft would be pretty inefficient; luckily, the British engineer Frank Whittle worked out how to solve this problem. He reckoned that since the sky is already full of gas, a plane shouldn't need to carry it around; you just have to compress the gas that's already in the sky as you fly along, and shoot it out the back. All you need is a machine to compress the air. This compressor is what you see under the wing as you board a plane—it looks like a giant fan, and it is, but what you can't see is that inside it are ten or more fans, each one smaller than the next. Their job is to suck in the air and compress it. From there, the compressed air goes to the combustion chamber, in the middle of the engine, where it's mixed with kerosene and ignited, producing a jet of hot gas that shoots out the back of the engine. The genius in the design is that, on its way out of the engine, some of the air's energy is used to rotate a set of turbines — and it's these turbines that rotate the compressors at the front of the engine. In other words, the engine harvests energy from the hot gas that it then uses to collect and compress more air as it flies through the sky.

The air shooting out the back of the engine allowed our plane, which weighed approximately 275 tons, to gain speed. It's always hard to get a feel for just how fast you're going when you're looking out the window of a speeding aircraft. The wings bob and wobble awkwardly at every bump of the runway, giving no hint of the engineering elegance that they'll display once airborne. At eighty miles per hour, the intensity of the rattling and groaning of the cabin interior begins a worrying crescendo. If I had never flown in a plane, at this point I'd be very doubtful that we would ever get off the ground.

And yet the sheer embodied energy in the kerosene propelled us forward faster and faster; a fuel with more power than nitroglycerin was being harnessed at a rate of one gallon per second. By now our aircraft was nearing the end of the 1.8-mile-long runway, traveling at 170 miles per hour. This is arguably the most dangerous moment of the flight. There wasn't much runway left, and if we didn't get airborne quickly, we would run off the end, plowing into the buildings there, with thousands of gallons of liquid kerosene in our fuel tanks. And yet majestically, like a goose taking off from a lake, the plane climbed into the sky, leaving behind all the buildings, cars, and people on the ground in a matter of seconds. This is the moment I love most about flying — especially when it involves flying through the low clouds of London into the bright sunshine above, as we did that day. It feels like entering another realm of existence, and I never tire of it.

A plane is, in a way, the modern magic lantern. Its genie is kerosene, which will grant your wish to go anywhere in the world, flying you there not on a magic carpet but in something even better, a cabin that protects you from the extreme cold and wind and is comfortable enough to allow you to relax, even sleep, during your journey.

Of course, like all genies, it has a dark side. We have fallen in love with the power of kerosene, but flying, and indeed the use of other products dependent on crude oil, are wreaking havoc with our global climate: it is warming rapidly as a result of carbon dioxide emissions from burning oils like kerosene. Globally we currently consume four billion gallons per day. Whether we will be clever enough to find a way of putting the genie back into the bottle is surely one of the most important questions of the twenty-first century.

But above the clouds I wasn't, if I'm honest, thinking about this. Instead I was marveling at the cloudscapes and looking forward to having a drink from the trolley, which, happily, was now trundling down the aisle.

2 / INTOXICATING

BY THE TIME WE were at cruising altitude—forty thousand feet—I was really enjoying myself, looking down from my window seat onto a layer of clouds, with the sun streaming over them into the cabin. I turned my head to the side and met my neighbor's gaze as she peered out the window too.

"It would be great to jump out now, wouldn't it, and dive into those big, fluffy, warm clouds," I said.

"They're not warm," she said.

"Er, no. You're right," I said. "I'm sorry."

Oh my God, did I really say that? I thought. Could it be the wine? Had it gone to my head already? I inspected the label of my miniature green-plastic bottle, which declared that the liquid I was drinking was a wine from Australia made from Chardonnay grapes. It was described as "full-bodied, with a vanilla buttery finish." I took a sip to see if I could taste the vanilla. I couldn't. There was acidity there, and something floral. I inspected the label again. The wine was 13 percent alcohol.

Alcohols are chemically similar to kerosene: for a start, they burn, as you will have witnessed if you've ever ordered a flambé dessert. Usually brandy is used for fancy dishes like that because this spirit has a high percentage of alcohol, typically 40 percent; and this is what burns with a bluish flame on top of your dessert.

Pure alcohol is easy to burn too and indeed is used as a fuel for cars. Brazil is the primary producer of alcohol made from sugarcane, which it uses as a transport fuel. The country is considered to have one of the most sustainable biofuel economies in the world, with some proportion of alcohol being used to fuel 94 percent of Brazilian passenger vehicles. This alcohol is made by turning sugarcane into juice and fermenting that with yeast. This is the same process by which both wine and beer are made: yeast consumes sugar and produces alcohol. But with biofuels the alcohol is then refined into pure alcohol. Biofuel isn't as popular in other parts of the world as it is in Brazil in part because other fossil fuels are much cheaper to produce, but also because it requires lots of land to produce crops that yield alcohol at the scale needed to sustain the transport systems of whole countries. So, worldwide, these crops are mostly grown to produce alcohol for drinking.

Alcohol is a key component of some of the world's most popular drinks, such as wine, beer, and spirits, but it is toxic. It is that toxicity that makes these drinks so intoxicating. That's where the word comes from. The toxins in alcohol suppress the nervous system, causing a loss of cognitive functions, loss of motor functions, and loss of control. It is quite surprising that despite these serious physiological effects, mild intoxication is so very enjoyable. In my case it causes me to become less uptight, to worry less, to grin; and, at higher doses, to dance badly without inhibition. Indeed, nothing quite hits the spot like an intoxicating drink at the end of a long week at work. "Drink me," a bottle of wine says, "and for a while the world won't be the same."

Alcohol is a general name for a family of hydrocarbon molecules similar to gasoline and diesel, but with an extra hydrogen atom and oxygen atom attached to them. Those extra atoms are called a hydroxyl group. Different kinds of alcohols come in different molecular sizes: the alcohol we drink has two carbon atoms and is called ethanol. It is a polar molecule, which means that there is a separation of a molecule's electric charge. In the case of alcohols, this is caused by the hydroxyl group. Water molecules also have a hydroxyl group and are also polar. This similarity is why ethanol dissolves in water. When the label on a bottle says what percentage

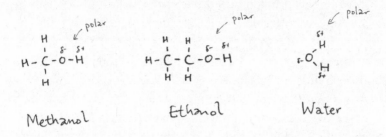

A comparison of the chemical structure of methanol and ethanol – both are alcohols. Methanol has one carbon atom, while ethanol has two. Both are polar molecules containing a hydroxyl group – the OH at the end. Water is also polar, and this similarity allows both methanol and ethanol to mix well with it.

of the drink is alcohol, it's telling you how much dissolved ethanol you're about to consume. In the case of the Chardonnay I was sipping, the answer was 13 percent.

Whereas one side of an alcohol molecule is similar to water, the other side, the hydrocarbon backbone, is similar to the structure of oils and the fatty molecules that coat the cells in your body. It is this similarity that allows ethanol to bypass the defenses of cell membranes and, being small, to sneak through the stomach's cell wall and enter your bloodstream directly. Approximately 20 percent of the ethanol you imbibe when you drink wine goes through your stomach wall and directly into your bloodstream, which is why you can feel the effects of alcohol almost immediately after drinking it.

This could explain my ludicrous remark to Susan, I thought, and quickly glanced her way to see if she seemed annoyed. Still wearing those red-rimmed glasses, she was engrossed in her novel. She had gray, close-cropped hair and wore a black T-shirt. Midfifties, I estimated her age to be. There were a few loose strands of hair, much longer than her own, on her T-shirt. *Were they her lover's hair,* I wondered, *deposited there as they hugged goodbye at the airport?* Or maybe they were from her dog.

Dogs also get drunk if they drink alcohol, which is why there is a growing market for non-alcoholic wine designed specifically for pets to consume at festive occasions. Non-alcoholic wine for human consumption is also available, although in my experience it bears very little resemblance to wine. What it does do, however, is highlight quite how much the regular wines rely on alcohol to balance the sweetness and fruitiness of the grape juice. It's what gives wine its air of sophistication and authority. Alcohol turns grape juice into an adult drink—a poison, admittedly, but one whose charms we willingly submit to.

I was already feeling a little intoxicated, but because I had not eaten anything for a while, I was about to become more so. Without food to slow down the progress of the alcohol through my stomach, it was now making its way to my small intestine. Here it entered my bloodstream and then encountered my liver. The liver's job is to get rid of the toxin, but it can metabolize ethanol at a rate of only about one glass of wine per hour (depending on your size). If you drink faster than this, ethanol will enter your bloodstream at a greater rate than it can be processed, and so will be able to infiltrate the rest of your organs, exerting its powers throughout your body. The effects of alcohol on the brain, for instance, are not uniform from person to person. They change, depending on how much you drink, your mental state, and other details of your physiology. But basically, alcohol depresses your nervous system, reduces inhibitions, and changes your mood.

Alcohol affects other organs too. It temporarily weakens the heart muscles, causing them to beat less vigorously and lowering your blood pressure. When blood circulates to your lungs to pick up oxygen from your breath, some of the alcohol jumps across the membranes along with the carbon dioxide being expelled from your blood. As you breathe out, the alcohol vapor becomes part of your exhale, which is why you can smell it when someone's been drinking. Testing for the presence of alcohol vapor in someone's breath is the principle behind the Breathalyzer, which the police use to test whether someone they suspect of drunk driving is, in fact, intoxicated.

While booze on the breath doesn't smell great, the other side of ethanol, the part that's more similar to oil than it is to water, gives us a considerably more fragrant liquid — perfume. Essential oils distilled from plants like bergamot and orange, or resins like myrrh, or animal-derived substances like musk, can all be dissolved in alcohol and turned into perfume. When you dab the perfume on your warm skin, the alcohol evaporates, leaving the oils to diffuse slowly into the air, shrouding you in the scent of your choosing. All the perfumes piled high in the departure lounges of airports are full of alcohol. If you were really desperate to get drunk, you could drink them; they'll have the same effect on you as vodka. But you have to be careful — some of the alcohols used in cheap perfumes contain methanol.

Methanol is the smallest alcohol molecule, with only one carbon atom, unlike ethanol, which has two. This small difference changes its pharmacological activity dramatically and makes methanol far more poisonous than ethanol. One shot glass of pure methanol can cause permanent blindness; three will kill you. This happens because once the methanol is in your body, your digestive system metabolizes it into formic acid and formaldehyde. Formic acid attacks nerve cells, especially the optic nerve. If you drink too much of it, the degradation of your optic nerve could leave you blind — this is where the expression "blind drunk" comes from. The formic acid also goes after your kidneys and liver, where it causes permanent damage that can be lethal.

Methanol is produced during the fermentation of alcoholic drinks, especially in the production of spirits like vodka and whiskey, but it's removed through the brewing process, so you're unlikely to encounter it in commercial spirits. If you make moonshine, hooch, poteen, or any other home-brewed spirits, though, you need to be very careful. These drinks are typically made by fermenting starch from crops such as corn, wheat, or potatoes.

This results in a low-alcohol mixture called a mash, which is then connected to some pipework called a still, heated up, and distilled into a liquor with a high percentage of alcohol. The first liquid that emerges from the still is concentrated in methanol—you have to throw it away. Experienced home brewers know this, but people die every year after making moonshine for the first time.

Those in search of cheap alcohol sometimes resort to drinking alcohol-based liquids that are easy to buy, such as antifreeze, cleaning products, and perfumes. This is a very bad idea—not just because these liquids taste foul, but also because, since they are not designed to be drunk, the manufacturers don't always remove the methanol they contain. This can lead to tragic consequences. For instance, in December 2016, fifty-eight people died in Russia by drinking a scented bath oil. It wasn't the scented chemicals that killed them, but the methanol.

Here on the plane, though, the drinks trolley was coming past again, carrying alcoholic beverages that I was confident had little or no methanol content. When the flight attendant got to us, she asked if we would like any drinks to accompany our meal. Susan asked for white wine, while I opted for red. "I couldn't taste the vanilla in the white," I said to her, "but see if you have any luck." Susan smiled, poured her wine, raised her glass to me, but said nothing and returned to reading her book. She seemed pleased that I had started to chill out, and so was I. Alcohol is, of course, a relaxant and a social lubricant—a drug, yes, but a legally sanctioned one that provides more benefits to society than the problems it causes—or at least that's the story we tell ourselves. Getting intoxicated can make people more relaxed, or it can make them more antagonistic. In either case, they also become less able to make clear, rational decisions. Which makes you wonder why the dangers of intoxication are not mentioned in the preflight safety briefing: surely a drunk person is less safe in an emergency, and less able to make good decisions that affect others? But then, that assumes that the briefing is really about safety, which, as already mentioned, I don't believe it is.

While drinking wine may not increase your safety, it has other uses, one of which the attendant had alluded to: it is a traditional accompaniment to meals. Apart from being delicious in itself, it acts as a very effective palate cleanser, making the food itself more enjoyable. One of the key flavor components of wine is its astringency: the feeling of dry, parched roughness in the mouth. Pomegranate, pickles, and unripe fruit are all astringent foods. In wines, the astringency comes from tannins. These molecules, which originate in grape skins, break down the lubricating proteins in saliva and leave you with a dry mouth. But still, mild astringency in drinks is pleasurable, especially when you're drinking them with fatty foods. Fats lubricate the mouth, but although they can make a dish feel rich and luxurious, in excess they mask flavor and coat your mouth in clagginess and sickly oiliness. Astringency counteracts this fatty feeling, cleaning the mouth, removing any aftertaste from the food, and resetting your palate to neutral.

Studies show that palate-cleansing works best when an astringent drink is sipped in between bites of fatty foods; the pairing keeps the dry-mouth feeling associated with high tannins from building up, and does the same for the slippery feeling of fattiness. In other words, it makes sense to drink a red wine with steak, or a fatty fish such as salmon, no matter what anyone says about drinking red wines with fish. People think red wine will overwhelm the delicate taste of the fish, which is why they advise white. But in fact the flavor profiles of white wines (fruity, vanilla, and so on) overlap with those of red wines, and so the blanket rule is not helpful. Really, it's much more important to consider a wine's acidity and sweetness as you choose one to accompany your meal. Acidity is a measure of the sourness of the drink, while sweetness is a measure of a wine's dryness in the mouth. Some people, for instance, prefer wines that balance the bitterness of food, so they'd want to pair their meal with a glass of something dry and acidic. For instance, a full-flavored white Rioja goes well with glazed ham, while red Pinot Noir works very well with a Mediterranean fish stew.

In many cultures, food isn't paired with wine, but with spirits like vodka. Spirits are very effective palate cleansers because they contain a high percentage of ethanol, often 40 percent, which provides astringency. The alcohol also dissolves oils and fats in the mouth, along with their associated tastes. The advantage of drinking pure spirits with food is that they have very little flavor and so will not clash with a strongly flavored dish such as pickled herring.

The reason pure vodkas have so little flavor is because they have very little smell. Although the basic tastes of salty, sweet, sour, umami, and bitter are detected by taste buds in your mouth, the complex flavor profiles of food and drinks are detected by the thousands of olfactory receptors in your nose. Hence the importance of the bouquet of wine — this is why wine enthusiasts always sniff before drinking; most of the flavor you taste really comes from the wine's scent. It is also why wineglasses are designed with a large bowl. This is a vessel designed to hold the bouquet of the wine for your delight and appreciation.

When you eat, the release of smells inside your mouth accounts for most of the food's flavor, which is why, when you have a cold and mucus is covering your smell receptors, you can't taste the subtleties of whatever dish you're consuming. It also explains why wine tastes different at different temperatures: when it's served cold, only the very volatile substances evaporate in your mouth, and so you experience the flavor profile dominated by those; but when you warm the wine up, the smell is different. The extra energy allows more of the flavor molecules in the liquid to evaporate. This changes the aroma of the wine and so its taste. One of the main reasons why red and white wines are perceived to taste so very different from each other is that they are served at different temperatures. Cool down both a red wine and white wine, and then drink them in a blind taste test, and you'll see what I mean. At cooler temperatures many of fruitier flavor molecules stay in the liquid rather than contributing to the bouquet. This changes the balance of the flavor, so that acidity and dryness are emphasized, and for many this imparts a sense of crispness and clarity. When combined with the cooling effect on the palate, this can be an extremely delightful experience — a classic white-wine moment. Serve the same wine at room temperature and it tastes completely different. Now the acidity is muted by a fruity, passionate embrace that's not crisp but rather warm. There is no right and wrong here — it is just a matter of what you enjoy.

The red wine I was drinking on the plane was probably at about 70°F; poured from a small bottle into my glass, the liquid had had time to warm to the ambient temperature of the aircraft.

Red wine in a glass, showing the Marangoni effect.

I swirled the wine around in the glass to gauge its alcohol content. I was looking for the Marangoni effect — when the wine forms tears as it flows down the glass. The ethanol in wine has the effect of lowering its surface tension with the glass, so when it's poured, it leaves a thin film. The alcohol in that coating quickly evaporates, leaving an area of liquid with a low concentration of alcohol, and thus a higher surface tension than the neighboring area. The unequal tensions pull the liquid apart, leaving a tear. The higher the alcohol concentration of the wine, the more pronounced this effect, so by looking at the Marangoni effect, you can get a sense of how alcoholic your wine is. My red had pronounced tears and so I estimated it to be a strong wine, with a high alcohol content of perhaps 14 percent, despite what the label said.

About that label. I closed my eyes and took a big glug without looking at the label to read the description. What could I taste? I found it had a strong, fruity, kind of, well, red-wine flavor. It wasn't bitter, but it wasn't sweet; it seemed fair to call it balanced. I wanted to say that it was smooth, but what did I mean by that? Clearly it was a liquid and therefore smooth by default. I guess I meant that it wasn't making my mouth feel dry or prickly — not astringent, then. *I liked it,* I thought, and allowed myself to peer at the label to see what it was meant to taste like.

"Deep violet, bags of blackcurrant and cherries, hints of bark; full of young tannins but still balanced, light body, fruity finish."

Ah ha! I thought, having a quick look at Susan to see if she was reading her book. Which she was. She looked up quizzically, and I realized I'd spoken out loud. This made me realize I was a getting a little drunk, but not so drunk that I wasn't capable of realizing it, which was good.

The taste of wine owes more to its appearance (especially the label) and its cultural associations than many wine experts would like to admit. Studies show that flavor is constructed in the brain, which takes inputs not just from the taste buds in the mouth and the sensors in the nose, but also from your brain's expectation of what things should taste like. For instance, if you take strawberry ice cream and use a flavorless dye to change its color, making it, say, green, yellow, or orange, then people who taste the ice cream will have difficulty detecting the strawberry flavor. More likely than not, they'll taste flavors related to the color. If the ice cream is orange, they're likely to taste peach; if it's yellow, vanilla; and often green will taste like lime. What's perhaps most extraordinary about this, though, is that when I tried it myself, even when I knew the orange-colored ice cream I was eating was strawberry, I still seemed to taste peach. Clearly, flavor is a multisensory experience, and as the brain constructs the taste of a food or drink using sensory inputs from multiple sources, sight is so dominant that it often overrides other sensory input.

There are many theories as to why flavor is so influenced by vision. One of the primary ones has to do with how our brain interprets fragrance. Flavor is constructed from smell, and our detection of smells is approximately ten times slower than our visual detection. We have great difficulty identifying odors from specific molecules. This might be because single odors are recognized by multiple receptors in the nose. Even experts trained to detect particular molecular substances through smell fail to do so when these are mixed with four or five other smells. When you consider that wine has thousands of individual flavor molecules, the staggering challenge of wine tasting becomes evident. That our sense of smell doesn't provide enough information to reliably distinguish between mixtures of odors is evident if you play a simple game. Blindfold your fellow dinner guests one evening, and ask them to identify the liquids in a series of glasses that you pass to them (try orange juice, milk, cold coffee). The rules of the game are that they may only smell the substances, not taste or see them. Some drinks are easy, but most are difficult for your senses to detect correctly. After this, do not reveal the answers, but instead allow your guests to take off their blindfolds and now use smell and sight to identify the substances. This is much easier, now that you can bring to bear your experience of seeing and smelling that particular drink in the past. The game illustrates just how much we rely on vision to identify smell, and thus taste.

The importance of vision in appreciating wine was demonstrated most dramatically in a scientific study carried out in 2001 in France. A panel of fifty-four tasters were asked to judge the bouquet of two wines and comment on them. Both were Bordeaux wines; one was a white, made from Semillon and Sauvignon grapes, and the other was red, made from Cabernet Sauvignon and Merlot grapes. But the participants didn't know that a flavorless red dye had been added to the white. So far as the participants could tell, they were smelling two glasses of red. The color completely dominated their appreciation of the bouquet of the wines. Participants described both wines using words like "spicy, "intense," and "blackcurrant," even though one was a white wine with a flavor profile that did not match these descriptors.

But no matter how we manipulate the color of our drink, when the flavor we taste matches what we expected based on the drink's appearance, we tend to enjoy it more. Similarly, the bottle from which it is poured, the cleanliness and ambience of the space we're in, the attractiveness of the person serving us, and — especially in the case of wine — the association of sophistication and quality all change our drinking experience. Experiments have shown that we'll like wine more or less depending on where the label says it was produced, and that we'll enjoy it more if we hear something good about it before drinking — that it's won an award, for instance. Rather a lot of wines win awards, by the way; in many competitions, the vast majority of wines entered by the producers win a commendation.

If you're one of those people who think they don't know anything about wine, and you feel bewildered when you're handed a wine list in a restaurant, think about the unfamiliar names of the grapes, the countries of origin, and the dates of production as you would specifications of a car. You may or may not care whether your car has a gas or a diesel engine, or whether it has a 1.4-liter engine or 2.0-liter engine. These details may not be something you want to learn about. You may just want a car to get you from A to B reliably, and that's really all that matters to you. Most midpriced wines will do this beautifully, the A to B in the case of wines being a pleasant accompaniment to food, or a vehicle to let alcohol shift your mood, or a way to celebrate a birthday.

But perhaps you're someone who likes a car to do more than take you from A to B. Maybe you enjoy the sensation of getting there, screaming fast around corners, for instance, or, alternatively, having a smooth, floaty ride. Some wines are a vehicle for spikier flavors than others, while some, such as "natural" wines, really push the boundaries of what you expect a wine to taste like. These are not better wines; they are different wines, because all taste is subjective, and as with cars (and most of life), price is no reliable guide to these experiences. When you enjoy a wine, just like a ride in a car, you are enjoying a multisensory experience. Equally, if you buy an expensive brand of car, that's really what you're paying for—the brand, not the experience. Some people love having the most expensive cars; they get real enjoyment out of what that says about them as a person. It's the same with wines. But this doesn't mean that these are better wines or better cars, or even that the owners are more sophisticated people. Thus, if having the most expensive wine doesn't turn you on, then you're wasting your money on $75 bottles of wine. Most midpriced wines and many budget-priced wines have flavor profiles that are just as complex as those of top-priced wines—and blind testing shows this.

Meanwhile, on the plane I had finished yet another glass of wine and felt a headache coming on. It couldn't be a hangover already, could it? Or was I just dehydrated? One of the physiological effects of alcohol on the body is to inhibit the secretion of hormones that tell your kidneys to conserve water. If you don't drink water to compensate, you become dehydrated. The cabin crew were nowhere to be seen, so I dug out the expensive bottle of water I'd bought in departures. The bottle hissed open and I took a greedy glug. That felt good. Looking out the window, I could see a much bigger body of liquid water below—the beautiful blue ocean, stretching all the way out to the horizon.

3 / DEEP

THE WATER IN MY plastic bottle was quite different from the water in the ocean I could see through my oval airplane window. The differences weren't just in composition — their respective salt contents and so on — but also in behavior. Earth's oceans are in a constant state of flux: they both create winds and are driven by them; they make the clouds and our weather systems, and are driven by them; they heat up the atmosphere, but also store heat. Huge global currents are established inside the oceans, and these affect our climate. Thus, despite being made of roughly the same molecules, the oceans that cover 70 percent of our planet are not just giant versions of the water in a bottle. They are utterly different beasts.

And *beast* is probably the right word to describe them. The oceans are dangerous, no matter how competent a swimmer you might be; keeping afloat in the open ocean is extremely difficult for more than a few hours at a time. My advice, if you do find yourself stranded at sea, is not to exhaust yourself trying to battle the currents; instead, float on your back while you await rescue. Though, in my opinion, *floating* is really the wrong word to describe what happens when humans bob about in the water. Floating is what boats do. They are majestic; they cruise along with just a small portion of their bulk submerged. Whenever I try to "float," most of my body sinks; if I'm lucky, I can just about keep my nose poking out of the water while I snort like a whale, breathing in air while simultaneously trying, and usually failing, to keep water from getting up my nose. Real floating, in my view, entails not just resting on top of water, but doing so with ease. But that's not the standard definition, and it's certainly not what Archimedes meant when he discovered the principle of floating two thousand years ago, and famously shouted "Eureka!" in his bathtub.

Archimedes was a Greek mathematician and engineer. He noticed that when you get into a bath, the water level goes up. The reason is obvious enough; you're sitting where some of the water used to be. It doesn't get compressed underneath you like a foam mattress would; instead, because it is a liquid, it flows around you and finds somewhere else to go. In the contained space of a tub, the only place for it to go is above the initial water level. If the bath is already full when you get in, then the water will flow over the edge of the tub and onto the floor. This is where Archimedes' famous experiment comes in. By collecting the water that spills over the edge in another vessel, it tells you something interesting: the weight of that water equals the so-called buoyancy force acting on you. If that force is lower than your weight, you will sink; otherwise you will float. This applies to any object. Eureka!

What Archimedes had discovered is that you can predict whether something will float or sink by simply working out the

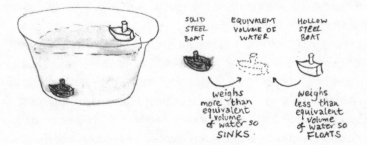

Why some things float and others sink comes down to whether they weigh more than their equivalent volume of water.

weight of water it will displace. For solid stuff you just have to compare the density of the material to the density of water. Thus wood, which weighs less per volume than water, is less dense than water, and so it floats. Steel is denser than water, so it sinks. But there is a trick: you can still make ships of steel if you make them hollow. Then their average density can be less than that of water, and so they float. It's as simple as that. Fast-forward two thousand years from Archimedes' heyday, and we find that the price of steel is now low enough for us to actually be able to build ships this way; our current maritime shipping fleet, which carries 90 percent of the world's traded goods, is made up almost entirely of steel ships.

The human body is composed of materials of varying density: there are dense bones and less dense tissue, and in some places we are hollow. Overall we're a bit less dense than water, which is why we can float. But if you adjust your density to exactly match the water's, by wearing something heavy—a metal belt, for instance —you'll be in a state of neither sinking nor floating; you're neutrally buoyant, the ideal state for scuba diving. When you're neutrally buoyant underwater, there's no net force trying to make you float to the surface, nor is there a force making you sink to the bottom of the ocean. In your scuba gear, you're effectively weightless, free to explore the coral reefs and sunken wrecks of the deep. It's so close to the feeling of weightlessness found in space that astronauts train in swimming pools.

Without the aid of scuba equipment, the human body floats. But our body is only slightly less dense than water, so more than 90 percent of it needs to be submerged in order to displace enough water to support our weight. Fatter people are more buoyant than thinner people because their fat-to-bone ratio makes them less dense. Wetsuits also make you more buoyant—they're coating you in a significant layer of material that is less dense than water. It's a little bit easier to float in the sea than it is in a swimming pool because the sea has minerals dissolved in it, such as salt, or sodium chloride. The sodium and the chlorine get inside the liquid by splitting up and inserting themselves between the water molecules. Having these atoms inside it makes the water more dense, so you don't need to displace quite so much water to counter your weight as you would in pure water. In fact, the Dead Sea in the Middle East has so much salt in it (ten times as much as the Atlantic Ocean) that you can bob around like a duck right on top of it.

A man floating on the Dead Sea.

Once you can float, you can swim: one of life's greatest pleasures. In water, not only are you weightless, but you can glide like a dancer. There's a hidden world lying just under the surface. Forget the expense of going to Mars and the excitement of searching for life on other planets — the oceans are, in all practical respects, alien worlds for us. By donning a pair of goggles and ducking underwater with a quick kick of the legs, we can visit them. Gliding down to the turquoise depths of a coral reef is one of the most wondrous things you can ever do. The fish observe you with weary eyes and flick their tails to swerve expertly out of your way. When you swim, you reach forward with one arm outstretched; by pulling it back, you cause the liquid around you to move rapidly enough to prevent the water molecules from moving past one another, so they jam together, exerting a force on you. It's that force that propels you forward in the opposite direction. This is the essence of swimming — your arms and your legs are constantly moving the water behind you, which has the effect of pushing you forward. It's not just thrilling; you essentially become a different person. Whereas on land you might be clumsy and plodding, in the water you can whirl and glide like a dolphin: you're free.

I used to live in a neighborhood in Dublin called Dun Laoghaire, within walking distance of a swimming spot called the Forty Foot, a rocky promontory in Dublin Bay famous for being featured in James Joyce's *Ulysses,* and home for centuries to a swimming club. I stopped by on a winter's day in 1999, and I saw people of all ages, but mostly senior citizens, jumping into the sea for a swim. The air temperature was perhaps 54°F, and the sea was about 50°F. I was wearing a big overcoat, and still I felt a bit cold as the winds from the Irish Sea buffeted me and the waves jumped up at the concrete quayside. Yet here were the elderly, who might be advised by their doctors to wrap up warm, jumping into the freezing waters. I got talking to a few of them as they dried themselves off afterward. They were happy and smiling and delighted. Their teeth chattered with the cold, but they were clearly elated. They told me that they swam every day of the year, in cold weather just as in warm — although, as I discovered in my time working there, Ireland rarely gets truly warm.

I decided to join them and bought a swimming cap that very same day. I swam in the Forty Foot every week of the year from then on. Looking back, it's one of the things I miss most about living in Dublin. But why did I love it so much?

Diving into water that is 50°F is not a comforting feeling. It is

The author after swimming at the Forty Foot in Dublin.

more like a slap in the face. It's not that the temperature is so extremely cold, but that you're surrounding your skin in water that's a good forty degrees colder than your skin. The water molecules draw heat away. But since liquids are denser than gases, there are many more molecules interacting with your skin per second than when you're just exposed to the air, so the heat conduction away from your warm skin is that much more extreme.

What makes it feel worse is another characteristic of water, called heat capacity. When water molecules are exposed to something hot, they jig about faster. These vibrations are what we call temperature. So the faster they go, the hotter the water gets. The hydrogen bonds holding the water molecules together strongly resist this vibration, so it takes a lot of heat to increase the average temperature of a quart of water molecules by even just one degree. To put this in perspective, it takes ten times more energy to heat up water than it does to heat up the same weight of copper. This feature of water, its exceptional heat capacity, explains why it takes so much heat to make a cup of tea. It also explains why an electric kettle is typically the most energy-intensive gadget in the kitchen. But that's just one of many ways in which water's high heat capacity — the highest of any liquid except ammonia — affects us. It's also what allows the oceans to store a lot of heat, and so their temperature always lags behind that of the air. Hence, on a sunny day in Dublin, the air temperature might heat up to 72°F, while the sea temperature will hardly change from 50°F. Sadly, for Irish people, this means that the sea never really warms up from the summer sun before winter comes again to cool it down. But it is a major advantage for us as a species because the high heat capacity of the oceans allows them to absorb a lot of the excess heat brought about by climate change. In other words, the oceans are stabilizing our climate, keeping us warm in the winter and cooling us down in the summer.

But none of that really explains why I enjoy swimming in the cold sea. I'm not one of those hardy outdoor types who relish being cold and wet. I'm a scientist and engineer, and I spend most of my time inside a laboratory or a workshop. Maybe that's the point — the sea is so marvelously wild and unpredictable that perhaps, unconsciously, I just want to expose myself to something utterly different from my day-to-day life. When you dive into the cold sea, you have to swim, to be alive and alert; it is so uncomfortable that it forces you out of a conscious rational mindset. It's impossible to worry about your failed experiments, unsubstantiated theories, or even your bungled relationships when you're gasping for breath — the very breath that is knocked out of you because you chose to dive into forbidding, uncontrollable waters.

Hypothermia is always at the back of your mind when you're swimming in cold water. Hypothermia sets in when your core temperature drops below 95°F. You start shivering uncontrollably and your skin changes color as your surface blood vessels contract, diverting blood toward your major organs. First you go pale, then your extremities turn blue. In very cold waters, the shock can cause uncontrolled rapid breathing, gasping, and a massive increase in heart rate that can lead to panic, confusion, and drowning. But even if you remain calm, swimming in 32°F water for just fifteen minutes will be fatal, as hypothermia sets in and shuts down your muscles.

Ultimately, I think it was that cold hand of death that drew me to the Forty Foot on all those icy gray January mornings, when the water temperature was, on average, 50°F. It made me feel more alive to be that close to death, to tease it, and then get out of the water and walk away unscathed.

Well, mostly unscathed. One day things didn't go so well for me. I arrived at the Forty Foot on a Saturday in February and found it deserted. The usual cadre of senior citizens was not in evidence. The tide was high and the water was choppy, with the occasional big wave coming in and launching itself onto the quay where I was changing into my swimming trunks. I was shivering, and my skin was covered in goose pimples from the cold winds. I was ready to jump, but hesitated before looking out at the water. I had never swum here alone before and the sea was rougher than I'd yet experienced. *Maybe,* I thought, *this is why no one else is swimming today?* Seconds of doubt elapsed. I remember goading myself, thinking, *Am I really so scared that, after going to the trouble of changing into my swimming trunks, I'm not even going to swim?* I dived in.

I felt the usual slap in the face, the sense that my body was under attack, that the ocean was sucking the life out of me. I always dealt with this by swimming energetically, so off I went, out to sea, fighting the waves coming toward me and trying to ignore the intense cold seeping into my limbs. I made it out a fair way before stopping to take a breather and getting hit smack in the face by a wave. I swallowed a mouthful of water, coughed, spluttered, and then took a deep breath, only to be hit in the face again. This time I choked. Water had gone down my windpipe, and I started thrashing out in an attempt to rise far enough out of the water to breathe properly, even if for just a couple of seconds. I couldn't do it, though; the water was too rough and the waves kept beating me down. I panicked, and began hyperventilating, while desperately kicking my legs to keep myself from drowning. Then another big wave hit me, and my panic turned to exhaustion. I couldn't win; I was cold and dead tired.

That's when I hit the rocks. While I'd been choking — for how long, I do not know — the waves and the tide had been pushing me toward the rocks that buttressed the Forty Foot and protected it from winter storms. These rocks, each the size of a small car, had been dropped into place by a crane to form a harbor barrier. Being washed up against rocks like that is normally something to avoid. It's nearly impossible to control the speed at which you'll hit them — that's almost entirely determined by the size, height, and speed of the waves carrying you — so it can be incredibly dangerous. But still, at that moment, I was relieved; hitting the rocks left me with a fair number of cuts and bruises, but it also gave me a chance to escape. Not that it was easy — as the waves that smashed me into the rocks receded, they pulled me away from the shore too. It took three or four waves and a good deal of scrapping, scratching, and bleeding for me to get a hold firm enough to allow me to climb up and finally escape from the sea.

I have relived this episode of my life many times, more often than not when I'm gazing out at the extreme implacable beauty of the ocean. But up here in the airplane, from my vantage point at forty thousand feet, the helplessness I'd felt at the time was magnified. I knew I could have drowned that day if I had swallowed one more wave, or if the tide had taken me out to sea rather than into the rocks. Many people die in similar circumstances. I knew I had been stupid. The ocean's ability to swallow you up without a trace is laid frighteningly bare when you look out at its harsh, seemingly endless expanse from the stratosphere. I turned to Susan, to see if she might be up for some sort of chat about oceans, waves, and accidental drowning, but she was wrapped up in a blanket with her knees pressed to her chest, watching some sort of sci-fi movie. The screen showed a picture of a spaceship moving into orbit around an enormous planet.

Size matters when it comes to bodies of water. When wind blows over a small pond, this creates friction, which slows the wind down and pushes against the water. This causes a depression in the water's surface. The surface tension of the water resists this change much as a rubber band resists being extended. Once that puff of wind ceases, just as with a rubber band, the release of tension, along with the force of gravity, restores the surface to its original form. As that water descends, a ripple is generated that radiates outward, as each water molecule displaces another, which, in turn, displaces still another, and so on. A ripple in water is really a pulse of energy. Energy, having originated from the wind, is now stuck on the surface of the pond. It makes the surface of the pond rougher, and so increases the resistance to the wind flowing over its surface. Thus, the ripple is joined by others, and they're pushed higher and higher. The higher the ripples, the greater the restoring force pulling them back down again, and so the rougher the pond becomes. There is a limit, though, to how high these ripples can go; eventually they'll hit the edge of the pond, and most of their energy will be absorbed by the land. But the longer they travel, the higher they'll get, which is why in a small pond ripples are never very big, but in a lake they can become so big that the wind will turn them into waves.

The top of a wave is called its peak and the bottom is called the trough. The distance between them is what we refer to when we talk about the size of a wave. As long as the size of the wave is smaller than the depth of the lake it's in, then the wave will travel uninhibitedly. But as the wave approaches the shallower waters at the shore, the trough will start to interact with the bottom of the lake, causing a kind of friction that will slow the wave down and force it to break, leaving it lapping on the beach.

In an ocean thousands of miles wide, those initial ripples have the time and space to grow to several feet in height. Wind blowing over the surface of the ocean for two hours at twelve miles per hour can make waves 12 inches high. A wind blowing at thirty miles per hour for a whole day can make waves 13 feet high. And a storm wind blowing for three or four days at forty-seven miles per hour can make 26-foot-high waves. The largest wave of this sort, recorded during a typhoon off the seas of Taiwan in 2007, reached 105 feet in height.

The waves produced during storms don't stop when the storm abates. Like ripples in a pond, they travel across the ocean, which is when their length becomes important. The length of a wave is the distance from its peak to the peak of the next wave. In a stormy ocean, it's hard to determine length because all the waves are jumbled on top of one another; a rough sea looks like a moving morass of angry water. When the storm ends, though, the waves carry on their way, and because they all have different lengths, they also have different speeds. So, as the waves travel across hundreds of miles of ocean, they separate out into sets, based on which ones are moving at similar speeds. Within the sets, the waves align, so that they run parallel. Eventually, each set will arrive at the coast in an ordered and regular pattern. Thus, the crash of waves onto the beach is essentially the sound of a storm that's come from very far away. That beautiful hypnotic rhythm comes from the complexities of ocean dynamics.

Given that storm waves are generated all over the ocean, it is a bit surprising that they usually approach land perpendicular to the beach. Surely, you might think, they should approach land at an angle determined by the straight line between the beach and whichever place in the ocean the waves were generated. But no, waves are too tricky for that. As a wave travels across deep water, its speed remains constant, because there's almost nothing that can slow it down. But as it approaches land, the water gets shallower, and the trough starts to interact with the seabed, slowing down that part of the wave. Meanwhile, the parts of the wave that have not yet encountered shallow water carry on at the same speed. The difference in speeds turns the wave in the same way that braking on one wheel of a car changes its direction. The net result: as waves approach land, they turn so that they are parallel to the contours of the seabed, which tend to run perpendicular to the beach, and thus most waves approach the shore from the same direction.

Surfers all know this. They also know about shoaling, which is what makes surfing such an exciting sport. Imagine you're sitting on your surfboard, looking out to sea; what you really want to know is where and when the waves are going to break. As the waves come closer to shore, they slow down because they encounter shallow water, but that also increases their height. This is shoaling. The shallower the water gets, the higher the wave gets, until the steepness of the wave reaches a critical angle where it becomes unstable. It's become so steep, you can slide down it on a surfboard, as if you were skiing down a mountain slope.

Surfing requires balance, timing, and an understanding of how waves behave. If you want to surf along a wave, you need part of the wave to start breaking before the rest. This means you need the contours of the seabed to slope gradually along the beach, because the moment a wave breaks is determined by the depth of the water it's moving through. You also need to understand the tides, which change the depth of the water throughout the day, based on the gravitational pull of the moon and the sun.

In sum, to catch a wave, you need a storm out at sea to produce waves big enough to travel across the ocean, toward a beach with an appropriately shaped seabed. You need them to arrive at just the right time of day to align with the tide. Then, if you're there at exactly that moment, wetsuit on, surfboard in hand, and ready, you might catch a sweet wave to shore. The exquisite timing of this confluence of events is what makes surfing such a special sport — it requires surfers to be completely in tune with the storms out at sea, the sun, the moon, and the water they're riding.

Even if you're not a wave connoisseur, it's still worth knowing about shoaling because it could save your life. On the morning of December 26, 2004, tourists on Phuket Island in Thailand were walking on the beach when they noticed something odd. The sea was receding fast, exposing usually submerged rocks and leaving boats stranded in the bay. Children watched and wondered, and so did their parents, as a large wave suddenly appeared; they thought they'd never seen anything like it before. But of course, they had: this was the shoaling of a wave. Only this wave was enormous: a tsunami.

As it turns out, just a few hours earlier, in the middle of the Indian Ocean, part of Earth's crust had ruptured, causing an earthquake of magnitude 9.0. This is a massive earthquake by any standards. The energy released was estimated to be ten thousand times bigger than that of the atomic bomb dropped on Hiroshima. Nevertheless, being that far out to sea, it didn't cause much immediate damage or loss of life. But the earthquake didn't just shear the tectonic plates of the crust—it also raised the seafloor by several feet. This, in turn, displaced approximately seven cubic miles of water. That's a lot of water—the equivalent of ten million Olympic swimming pools. And just as moving suddenly in the bath makes the water slosh back and forth, the earthquake set this enormous amount of water into motion.

Waves being waves, they set off across the ocean in all directions. If you'd been looking down from an aircraft at the moment the tsunami began, you probably wouldn't have been too worried. The waves were spread over such a large distance, and in such deep water, that only a small hump would have been discernible. But you might still have been alarmed by the speed at which they were traveling. Because of the intensity of the earthquake, and the vast amount of energy released over a short period of time, those waves traveled at the speed of a jet aircraft, about five hundred miles per hour. As they approached the coast and the shallow water of the Andaman Sea, they slowed down and got taller.

The closer they got, the more the shoaling intensified. Because the waves were hundreds of feet long, the first thing the people on the beach noticed was the water being sucked out to sea. If they had recognized the phenomenon, they would have had about a minute to run to higher ground. But, tragically, most of them didn't know what was happening—unlike many of the animals near the beach, which seemed to sense that something strange was going on, and fled. Those who stayed were hit by the first wave, which was thirty feet high when it reached the shore.

The arrival of a tsunami wave.

All in all, the tsunami killed 227,898 people along the coastlines of fifteen countries. What makes a tsunami so dangerous is not just the vast amount of water it dumps on the coast, but the force that the water exerts on everything it meets. One gallon of water weighs 8.3 pounds, and the tsunami displaced eight trillion gallons of water. It ripped apart huts, trees, and cars, destroying them and thus creating a river of debris that smashed into everything it encountered. It swept up tankers, houses, and cars and flung them into bridges and overhead electricity pylons, which collapsed, creating lethal fires. The people who were pulled into the wave were carried along, bashed, tumbled, and crushed by all this fast-flowing debris. This knocked many of them unconscious or injured them in a way that hindered their ability to stay afloat. Just like storm waves, tsunamis come in sets, and as the first wave was pulled back (having reached more than a mile inland in places) by the approach of the second, the currents reversed and pulled the people and debris caught in their path into this new onslaught.

Unfortunately, those who were lucky enough to survive this devastation faced any number of challenges in its aftermath, water pollution being one of the most severe. The freshwater supplies in the areas hit by the tsunami had been poisoned by the destruction of sewers and the infiltration of saltwater; the hundreds of thousands of people struck dead by the waves all had to be buried as quickly as possible to prevent the spread of disease and pests; and longer-term infiltration of saltwater into the region's arable land left it unable to support crops.

But as catastrophic as the 2004 tsunami was, the one in 2011 off the coast of Japan was even more powerful. The tsunami was created by the force of an enormous earthquake—the fourth most powerful in recorded history—with an epicenter in the ocean, forty-three miles off the shore of Honshu, the largest island in the Japanese archipelago. Shaking was felt on land for six minutes, but the worst damage didn't occur until later, when the resulting tsunami hit the shore, devastating entire towns and colliding with the Fukushima Daiichi Nuclear Power Plant.

Fukushima Daiichi was built in 1971 and had six nuclear fission reactors. These reactors are made up of rods of uranium oxide, which are bundled together inside the reactor core. A reactor emits radiation in the form of very high energy particles. In a nuclear power plant, most of this energy goes toward heating up water to create steam, which drives turbines, which in turn create electricity. This type of nuclear energy is so powerful that a set of rods of uranium oxide the size of a small car will produce the amount of electricity needed to run a city of a million people for two years. Prior to the 2011 tsunami, the Fukushima plant had six of these reactors, all producing power twenty-four hours a day, 365 days a year, for approximately five million people.

Japan has a long history of earthquakes; it lies at the boundary between two major tectonic plates. The Fukushima plant was built to withstand these earthquakes, and indeed it did. As did Japan's other fifty-four nuclear reactors. When the earthquake occurred, on March 11, 2011, it didn't damage the plant at all. However, due to legally mandated safety precautions, three of the reactors (1, 2, and 3) all shut themselves down (reactors 4, 5, and 6 were already shut down for refueling). You can't just turn nuclear fuel "off." It still gives out heat and radioactivity when the reactors are shut down. They need active cooling to prevent a meltdown of the uranium oxide. During shutdown, this is provided by diesel-fueled backup generators, which produce electricity to power the pumps that circulate the cooling water.

Ultimately, thirteen thousand people would die as a result of the 2011 earthquake; but when the shaking stopped, and the reactors shut down, 90 percent of them were still alive. Then, fifty minutes later, a forty-three-foot tsunami wave, traveling at an average speed of 310 miles per hour, hit the power station. The water destroyed the plant's seawall defenses and flooded the buildings containing the diesel generators that were cooling the nuclear fuel rods. The generators failed, and a second backup system kicked in, powered by a set of electrical batteries. The batteries had the capacity to run the plant's cooling systems for twenty-four hours. Under normal circumstances, this would have been enough time either to restore the diesel generators or to obtain more batteries. However, the tsunami, the biggest to hit Japan in modern times, destroyed anything and everything in its path. The sheer force of the water pulverized whole towns, forty-five thousand buildings, and almost a quarter of a million vehicles, and left the regions' roads and bridges in a mess. The areas where the tsunami hit came to a standstill, making it incredibly difficult to get medical help to the survivors and impossible to rush the backup batteries to the Fukushima plant in time to replace the ones running the cooling systems. Twenty-four hours after the tsunami hit, the batteries died and the temperature inside the reactors started to rise.

When nuclear fuel rods melt, they look a lot like lava, but the liquid is much hotter. Lava comes out of a volcano red-hot, typically at 1,800°F. Liquid uranium-oxide nuclear fuel is much more fearsome, a white-hot liquid with a temperature exceeding 5,000°F. It will melt and dissolve pretty much anything it comes in contact with. At Fukushima, it melted its way through the ten inches of steel that had been containing it, and then continued to eat through the concrete floor of at least one of the reactors. But that was just the beginning.

The nuclear fuel in the reactor is encased in an alloy made from zirconium. It is incredibly resistant to corrosion, except at high temperatures. At 5,000°F, the zirconium alloys react strongly with water, producing hydrogen gas. It's estimated that as a result of the meltdown, one ton of hydrogen gas was produced in each of the plant's reactors. On March 12, the hydrogen gas reacted with the air inside the reactor containment building, creating an explosion that destroyed the complex.

Liquids are incredibly hard to contain, and as a result a great deal of the radioactive contamination from these nuclear meltdowns made its way into the area's water systems, and ultimately the sea. From there it can and does go anywhere and everywhere. This is why the primary concern of all nuclear-waste engineers is preventing water ingress into any of their storage facilities. Yet most nuclear power stations are built near large bodies of water, not because it's safer, but because it's cheaper. They need the water for cooling: having a large supply readily available makes the plant much more energy- and cost-efficient. But as we saw in Fukushima, when disaster strikes, our water supply is vulnerable to a huge amount of radioactive waste.

This isn't just a nuclear issue. Almost all of the world's major cities are coastal because, historically, trade between countries required ports. But with sea levels rising as a result of global climate change, the impact of tsunamis and hurricanes and storms is going to make these places—and their dense populations—ever more vulnerable. The only way to protect ourselves from this threat is to get to higher ground, or perhaps into the air. A tempting thought, from my perch on the plane, where I was currently sipping water and looking down omnipotently at the vast Atlantic. It was a calm, clear day, and the ocean, I felt, looked almost innocent.

But then there was a bump, and the whole aircraft seemed to drop for a second before recovering. Then it happened again, so violently that water shot out of the top of my bottle and soaked my lap.

"We are currently flying through a patch of turbulence," the captain announced over the intercom. "I'm turning on the FASTEN SEAT BELT sign, and ask that all passengers please return to their seats. We will resume the in-flight service in a few minutes when we reach smoother air." The plane dropped vertiginously again. My stomach felt queasy, and out the window I caught a glimpse of the wings oscillating wildly.

4 / STICKY

IT DOESN'T MATTER HOW many times I experience turbulence on a plane; I can never seem to stop the seeds of panic forming in my brain. Rationally, I knew the wings weren't going to snap — we were flying in one of the most technologically advanced passenger aircraft ever made. I've even visited the factory where the wings are glued together and seen them being mechanically tested. But despite this, the rational part of my brain was being ignored by my panicking neurons. I know I'm not alone in this. Over the years, I've learned not to talk about the aircraft being glued together to other passengers; they generally don't find that reassuring.

Many liquids are sticky — that is, they will stick to you if you put your finger in them. Oil sticks to us, water sticks to us, soup sticks to us, honey sticks to us. Thankfully, they stick to other things better than to us, which is why towels work. When you have a shower, water trickles down your body, sticking to your skin instead of bouncing off, allowing it to follow the curves of your chest, belly, and bum in defiance of gravity. This stickiness is due to the low surface tension between the water and your skin. When the water comes in contact with the fibers of a towel, they act as little tiny wicks — just as candle wicks suck up liquid wax — so the micro-wicks of the towel suck the water off your body. Hence, your skin gets dry, and the towel gets wet. The stickiness of liquids, then, is not a property intrinsic to any particular liquid. It is determined by how liquids interact with a particular material.

However, just because something is sticky doesn't mean it can be used to glue an aircraft together. Wet your finger and dab it on a speck of dust, and it will stick to you, and remain sticking to you until the water evaporates. That water loses its stickiness when it evaporates is why, although it is sticky, it can't be considered a glue. Glues start off as liquid and then, generally speaking, turn into a solid, creating a permanent bond.

This is a materials process that humans have been playing with for a very long time. Our prehistoric ancestors made pigments from the powder of charcoal or naturally occurring colored rocks like red ocher, and used them to draw pictures on the walls of

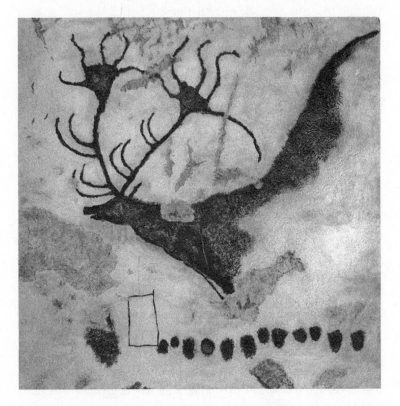

An ancient charcoal-and-ocher cave painting of *Megaloceros*, an extinct genus of deer.

caves. To get the pigments to stick to the walls, they mixed them with sticky things like fats, wax, and egg, and so invented paints. Paints are essentially colored glues, and these earliest ones were permanent enough to last for thousands of years. Some of the oldest cave paintings still in existence are in the Lascaux caves in France, estimated to be about twenty thousand years old.

Tribal cultures have long used these colored sticky substances as face paints, a central part of both sacred rituals and warfare. The tradition continues today with the modern cosmetics industry. Lipstick, for instance, is made up of pigments mixed with oils and fats that allow the color to stick to your lips — hence the name. Getting the glue to stick to your lips for hours, but still be easy enough to remove at the end of the day, has always been an issue; ditto for eyeliner and any other kind of makeup. The problem illustrates one of the main themes in glue design — namely, that unsticking is often as important as sticking. But more on that later — mastering sticking is hard enough for now. If you want to stick something together that needs mechanical strength, like the components of an ax, a boat, or indeed an airplane, then you need something stronger than paint or lipstick.

In the summer of 1991, two German tourists discovered the mummified remains of a man while walking in the Italian Alps. He turned out to be five thousand years old and was later nicknamed Ötzi. His remains were extremely well preserved because they'd been encased in ice since his death, as were his clothes and tools: he wore a cloak made of woven grass, and a coat, a belt, leggings, a loincloth, and shoes, all made of leather. All of his tools were ingeniously designed, but with respect to glue, it's Ötzi's ax that's the most interesting. Made of yew wood, with a copper blade, the ax was bound together with leather straps stuck on with a birch resin. This gummy substance is produced by heating up birch bark in a pot, yielding a brownish-black goo that was widely used as an adhesive in the late Paleolithic and Mesolithic periods. It works for making heavy tools like an ax because when it cools, it forms a tough solid. Our ancestors used it

to attach an arrowhead to a shaft, to make flint knives, to repair pot-
tery, and to make boats. The liquid is mostly made from a family of
molecules called phenols.

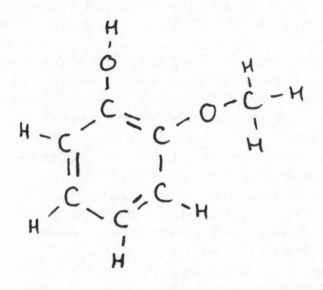

The molecular structure of 2-methoxy-4-methylphenol, which is one of the constituents of birch-bark glue. The hexagon of carbon and hydrogen atoms joined
to an –OH hydroxyl group is the hallmark of a phenol.

The chemical name may be unfamiliar, but I'm sure you would
recognize the smell: the major phenol in birch-bark glue is 2-me-
thoxy-4-methylphenol, which smells of smoky creosote. Phenol
aldehyde smells of vanilla. Ethyl phenol smells of smoky bacon; indeed, whenever you smoke fish or meat, it's the phenols that give it
that distinctive flavor.

When you heat up the birch bark, you extract the phenols. The
thick resin that's produced is basically a mixture of a solvent called
turpentine and phenols. The turpentine is the base of the liquid, but
over the course of a few weeks the turpentine evaporates. This leaves

just the phenol mixture, which turns from a liquid into a hard tar sticky enough to bond wood to leather and other materials.

As it turns out, trees are excellent purveyors of sticky things. Pine trees exude nodules of resin that also make good glues. An adhesive popular for a thousand years, gum arabic, comes from the acacia gum tree. The resin of trees of the genus *Boswellia* is a particularly nice-smelling glue called frankincense. Myrrh, another aromatic resin, comes from a thorny tree of the genus *Commiphora*. Resins were often used in medicines as well as in perfumes, perhaps because their active chemical components, like phenols, had potent antibacterial properties. Frankincense and myrrh were so highly valued in antiquity that they were given as presents to queens, kings, and emperors, which is why their presence in the Christian nativity story is so significant.

The stickiness of tree resins is no accident. They evolved to be sticky so they could trap insects, and therefore provide a valuable form of defense for the trees. The gemstone amber is actually fossilized tree resin, and often there are insects and bits of debris trapped inside it, perfectly preserved.

Without tree resins, it would have been very hard for our earliest ancestors to make tools and equipment, and so for our civilization to get off the ground. Nevertheless, you wouldn't want to glue an airplane together with them—they would certainly crack during flight. Phenol molecules don't bond very strongly to other substances—the molecule itself is too self-contained, too happy with sticking to itself.

But once you're in the trees, you don't need to look very far for stronger glues. Consider birds: their wings are not bolted or screwed together. Their muscles and ligaments and skin are bonded via families of molecules called proteins. Our bodies are joined together with them too. One of the most important of these proteins is called collagen. It is common to all animals and relatively easy to extract. Early humans used skins of fish and hides of wild game—they separated the fat and then boiled the skins

An ant trapped in amber, a fossilized tree resin.

in water. This extracts the collagen from the animals and creates a thick, clear liquid that turns into a solid, stiff material when it cools: gelatin.

The collagen proteins in gelatin are long molecules made from a carbon-and-nitrogen backbone. In animals, collagen molecules stick together to create strong fibrils that make up tendons, skin, muscles, and cartilage. But once they've reacted with hot water in the glue-making process, the collagen molecules separate. They now have chemical bonds to spare, which they want to satisfy. In other words, they want to stick to something else — they've become the animal glue gelatin.

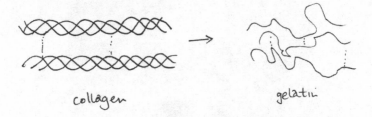

collagen   gelatin

How the structure of collagen fibril is transformed to become the animal glue gelatin.

It was animal glues that replaced wood resins as the mainstay of early human technologies. The Egyptians, for instance, used animal glue to make furniture and decorative inlays. In fact, it appears that the Egyptians were the first people to use glue to get around one of the main mechanical problems of working with wood — that it has a grain.

The density and arrangement of the cellulose fibers in wood give it its grain, which is determined not just by the biology of trees but also their growth environment. Thus the grain varies from species to species and from tree to tree. The upshot: wood is strong *across* the grain but has a tendency to crack *along* the grain. This is not a big deal if you are splitting logs for a fire, but if you are building a house, a chair, a violin, an airplane, or pretty much anything out of wood, it presents a design problem. The thinner the piece of wood, the more likely that it will crack. Counterintuitively, the solution to this problem is to cut the wood into even thinner pieces, called veneer.

The Egyptians were the first to make veneer. They stuck thin pieces of it on top of one another, so that the grain of each layer was perpendicular to the one above. This structure didn't tend to crack in any direction: we now call it plywood. The Egyptians used animal glues to stick the plywood together, and that worked reasonably well. But if you've ever cooked with gelatin, you know that animal glue dissolves in hot water. Unless kept absolutely dry, furniture made with animal glues falls apart. This seems like a huge defect, but Egypt is and was a very dry place, and so they managed.

And as mentioned earlier, there are in fact distinct advantages to a glue that can be unstuck. Historically, the designers of classic musical instruments, like Antonio Stradivari, known as the greatest violin maker of all time, used animal glue to construct their instruments. This would have allowed Stradivari to unstick any faulty joints during production, and so build almost perfect instruments. Today, in order to repair a wooden instrument, craftspeople unstick the joints with steam. This causes the bond between the glue and the wood to weaken and then dissolve. Thus, the wood comes away undamaged and clean, extending the life of the instrument and increasing its value. Indeed, most people who work in furniture restoration use animal glues precisely because they can be easily unstuck by using heat.

But when it comes to making wings, heat can be a real problem, or at least that's what legend tells us. Just look at what happened to the mythical king Minos, who ruled the Mediterranean island of Crete. The sea god Poseidon gave the king a very beautiful snow-white bull. King Minos was instructed to sacrifice the bull to honor Poseidon, but he sacrificed a different one instead because he did not want to kill the more beautiful bull. To punish him, Poseidon made King Minos's wife fall in love with the bull, and the offspring of that union was a creature that was half man and half bull—a Minotaur. This Minotaur grew up to be a terrifying beast that ate humans, and so King Minos got his master craftsman Daedalus to construct a prison for the Minotaur, in the form of an elaborate maze called the Labyrinth. To prevent Daedalus from telling others about the secrets within this structure, King Minos imprisoned him in a tower, along with his young son, Icarus. Daedalus, though, was a hard man to contain. He constructed wings by gluing feathers together with wax: one pair for himself and one for Icarus. On the day of their escape Daedalus warned his son not to fly too close to the sun. But during their flight Icarus was so exhilarated that he began to soar higher and higher. The wax melted, the feathers came unstuck, and Icarus fell to his death.

The fall of Icarus, which perpetuates the myth that he fell because the wax holding his wings together melted.

If you are wondering whether a modern aircraft could come unglued as it flies higher and higher, I should point out that the myth of Icarus defies logic. By flying higher, Icarus would have experienced colder, not hotter, temperatures. Temperature decreases by 1.8°F for every thousand feet of altitude you gain because the atmosphere is cooled by the radiation of heat into space. At forty thousand feet, the altitude my plane was flying at, the temperature outside my window was approximately $-60°F$, a temperature at which all waxes remain solid.

I should also say at this point that modern aircraft are not glued together with wax — nowadays we have much better glues. The intellectual journey of how they were discovered starts with rubber. Rubber is, of course, another sticky tree product. It's extracted by tapping the bark of the Pará rubber tree, which is indigenous to South and Central America. Mesoamerican cultures made many things with it, including the bouncing balls they used in ritualistic games. When European explorers reached the continent in the sixteenth century, they were amazed by rubber. They had never seen anything like it: it has the softness and pliability of leather, but is far more elastic and completely resistant to water. But despite its obvious value, no one in Europe could find an immediate economically viable use for it. Then the British scientist Joseph Priestley found that it was good for *rubbing* pencil marks off paper — which is how rubber got its name.

Natural rubber consists of thousands of small isoprene molecules bonded together in a long chain. This molecular trick of linking together units of the same chemical to make a completely different one is common in nature. These types of molecules are called polymers — *poly*, meaning "many," and *mer*, meaning "unit." Isoprene is the *mer* in natural rubber. The long polyisoprene chains in rubber are all jumbled up like spaghetti. The bonds within each chain are weak, which is why there is not much resistance if you pull rubber: the chains just unravel. This is what makes rubber so stretchy.

It's rubber's stretchiness that makes it so sticky. It can be molded easily and wedged into any space, including the crevices in your hand; this stretchiness makes it grippy. Therefore rubber is the perfect covering for the handlebars of a bicycle: it prevents your hands from accidentally slipping off, but you don't need to worry about getting stuck to the bike forever. The same goes for making

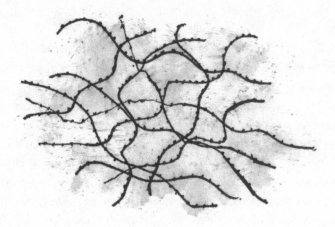

The structure of natural rubber, which consists of a jumble of long polyisoprene molecules.

car tires—rubber lets the car adhere to the road strongly enough to create the friction needed to move the wheels forward, but not so strongly that the car gets stuck to the road permanently.

One little-known but especially ingenious use of rubber is on Post-it notes. Post-its have an adhesive layer of rubber that remains stuck to the notes when you pull them from their pad, so they can be attached to walls, tables, computer monitors, books, and more, without damaging these surfaces or leaving a mark. The microscopic spheres of rubber that make up the glue on the Post-it bond strongly to the note itself, but when pressed onto a surface, they create only a small adhesive force. Which is why, when you pull the Post-it off whatever it's stuck to, the rubber stays put on the paper. Thus, the Post-it is repositionable and reusable. Genius? Well, actually, this not-very-sticky glue was in fact an accidental invention. In 1968 Dr. Spencer Silver, a chemist with the 3M Company, stumbled upon it while he was trying to make a super-strong adhesive.

Many other culture-changing adhesive products emerged in the twentieth century. One of the most important was sticky tape, invented in 1925 by another 3M Company inventor, Richard Drew. Drew's tape is composed of three key layers. The middle layer is made of cellophane, a plastic made from wood pulp that gives the tape its mechanical strength and transparency. The bottom layer is an adhesive, and the top layer — the crucial layer — is a nonstick material such as Teflon, which has a high surface tension with most other materials and so cannot be easily wetted by them (which is why we use it in nonstick pans). Its use in tape really is genius; it means the tape can be placed on top of itself without permanently sticking to itself, allowing it to be manufactured as a roll. And a roll of tape — well, what home is complete without one? — or ten, in my case.

You can tell a lot about someone from how they handle a roll of sticky tape. I have to admit straightaway that I'm a tearer, not a cutter. Ask me for a bit of tape, and I'll grab the roll and enthusiastically try to rip off a piece for you. I probably won't get it right the first time. Most likely I'll mangle a few pieces first, either tearing them off at a crazy angle or snapping them off clean, and inevitably I will somehow allow the sticky parts to get stuck to one another. I'm not proud of this; it actually sends me into a rage. I get increasingly furious with the tape, which in turn seems to goad me by sticking itself back onto the roll so seamlessly that I can't find the end. At this point I have to resort to running my thumb around the roll, trying to locate the end by feel alone. This sometimes takes so long that I start shouting at the tape. Then I throw it across the room — and wonder why it is that I still don't own a tape dispenser.

Gaffer tape suits my personality better. It's designed to tear without scissors. It's reinforced with fabric that runs across the roll, and makes the easy tear possible. The strength of the tape comes from the fabric's fibers, while the stickiness and flexibility come from the plastic and adhesive layers. I love gaffer tape so much, I confess I envy people with jobs that require them to carry it around with them on their belt. Thinking of this, I snuck a glance at Susan, who was still watching a film, and wondered what kind of tape she might favor. Her book, *The Picture of Dorian Gray* by Oscar Wilde, was lying on the tray table in front of her. I noticed the spine had been repaired with what looked like red electrical tape. The ends of the tape were clearly cut with scissors: so she was that kind of person.

The sticky tape pioneered by Richard Drew, although a useful invention, is not the technological innovation that led to the modern aircraft. That came from another American chemist, Leo Baekeland, who succeeded in making one of the first plastics. He made his plastic by combining two liquids. The first was based on phenols, the main constituent of birch resin, and the other ingredient was formaldehyde, an embalming fluid. These two liquids react together to produce a new molecule that has a spare bond for more phenols to attach themselves to, which in turn produces bonds for still more reactions with more phenols. Eventually the whole liquid (if you get the proportions right) is chemically locked together into a solid. In other words, the reaction makes one giant molecule, and all the bonds holding it together are permanent, so whatever object you've created will be hard and strong.

Baekeland used this new plastic to create a number of objects such as telephones, which had just been invented. This was, of course, immensely useful and made Baekeland a fortune. But it had another impact. Chemists realized that the phenol and formaldehyde could be mixed and applied to the interface between two things — gluing them together as it hardened. This was the beginning of a whole new family of glues called two-part adhesives, which were stronger than anything that had come before.

The more people used these two-part adhesives, the more they understood just how useful they were. First of all, the different components — phenol and formaldehyde — could be stored in separate containers, and thus remain liquid until they needed to be used.

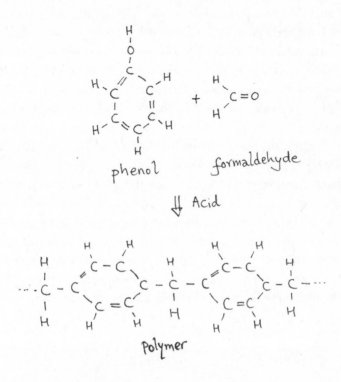

How two liquids, phenol and formaldehyde, create a strong adhesive.

And then, beyond that, you could alter their chemical composition through additives, and make them better or worse at wetting and then sticking to different materials, such as metals or wood.

This new type of glue had a big effect on the world of engineers. They returned to thinking about plywood, first developed in ancient Egypt. If you made plywood using a two-part adhesive perfectly designed to bond with wood, you'd have plywood that was neither held together with the weak bonds of animal glue nor sensitive to water. But for this new plywood to take off, a strong market demand was needed. The simultaneous development of the aircraft industry provided just that. In the early twentieth century, most planes were made of wood, but because of wood's grain, they were liable to crack. Plywood was the perfect solution — it could be molded into aerodynamic shapes and, thanks to the new two-part adhesives, was both reliable and resilient.

The most famous plywood airplane ever built was the de Havilland Mosquito bomber. When it was introduced during World War II, it was the fastest aircraft in the sky. Because it could outrun every other plane, it wasn't even outfitted with defensive machine guns. It remains, to this day, perhaps the most beautiful plywood object ever made. Its elegance and sensuousness come from plywood's ability to be molded into complex shapes while its glues set, a property that made it popular among designers for decades.

After the war, plywood continued to revolutionize our world — this time with furniture. Two of the most innovative designers at the time were the married couple Charles and Ray Eames, who

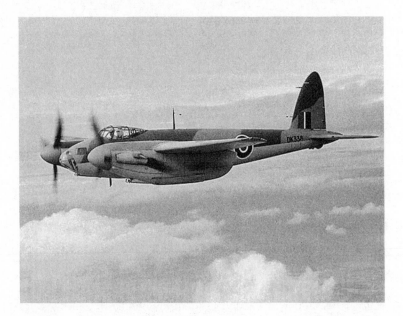

A photo of a de Havilland Mosquito bomber, which was made of plywood.

used plywood to reimagine wooden furniture. Their designs, particularly what is now known as the Eames chair, became classics, and are still made and imitated today. Go into any café or classroom and you are likely to see them. Other fashions in furniture have come and gone, but plywood has retained its appeal.

A plywood chair designed by Charles and Ray Eames.

But while plywood furniture stood the test of time, aeronautical engineering had to move on. After the war, aluminum alloys became the preeminent material for making aircraft, not because they were stronger by weight than plywood, or even stiffer by weight than plywood. No, aluminum won out because it could be more reliably manufactured, pressurized, and certified, especially as planes became bigger and started to fly higher. It's very hard to stop plywood from absorbing water or from drying out. Plywood aircraft that spent a lot of time in arid countries would eventually dry out, causing the material to shrink and putting stress on the glued joints. Similarly, for aircraft deployed in very wet places, the plywood would expand (or even rot), again compromising the safety of the aircraft.

Aluminum doesn't suffer from these defects; in fact, it's incredibly resistant to corrosion and, as such, formed the basis of aircraft structures for the next fifty years. But it is by no means perfect — it isn't stiff enough or strong enough by weight to create truly lightweight, fuel-efficient aircraft. So even when aluminum aircraft production was at its height, a generation of engineers kept scratching their heads, wondering what the ideal material for the skin of an airplane might be — another metal, or something else entirely? Carbon fiber looked promising since it was ten times stiffer by weight than steel, aluminum, or plywood. But carbon fiber is a textile, and at the time, no one could make a plane wing out of it.

The answer, it turned out, was epoxy glue. Epoxies are another two-part adhesive formulation, but at their core is always a single molecule called an epoxide.

There is a ring at the center of the epoxide molecule, with two

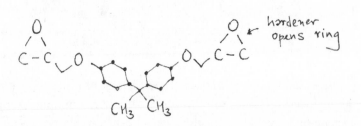

The ring of an epoxide molecule being opened up by a hardener, allowing it to form a polymer glue.

carbon atoms connected to one oxygen atom. Breaking these bonds opens up the ring, allowing the epoxide to react with other molecules to create a strong solid. The hardening reaction won't start until the ring is opened up by breaking the carbon-oxygen bonds, and this is typically done by adding a "hardener."

One of the major advantages of epoxies is that the reaction is temperature-dependent; you can mix it up and it won't start bonding until you want it to. This is critical to the production of the complex-shaped, fiber-reinforced parts that make up an airplane wing; the parts are all enormous and require weeks to manufacture. When you are finally ready to transform the glue into a strong solid, you put it into a pressure oven, heat the wing up to the right temperature, and presto.

These ovens are called autoclaves, and they can be the size of aircraft. All air is removed from the aircraft molds before they are heated up, solving another problem related to glues—air often becomes trapped inside a bond, forming a bubble that, once it has hardened, becomes a weak spot. Another major advantage of epoxides is that they are chemically very versatile. Chemists can attach different components to the epoxide ring, which allows it to bond to different materials, such as metals, ceramics, and, yes, carbon fiber.

You may be wondering why the epoxy resins sold in hardware stores don't need to be heated and autoclaved before you can use them to repair broken crockery or glue the knob back onto the metal lid of your juicer. These epoxies have chemical hardeners different from the ones used to make aircraft, and they've been designed to react with the epoxide molecule at room temperature. The glue is sold in two containers, which you have to mix together. One tube holds the epoxide resin and the other tube contains a hardener and various accelerators that speed up the reaction, allowing the glue to become solid faster. These domestic epoxies are not as strong as the aerospace versions, but they are still very powerful.

Maybe this all sounds easy, but it took decades to develop the

fundamental understanding and technology of composite structures to the point where everyone would trust these carbon fiber
planes in flight. First, carbon fiber composites were tested on the
ground in racing cars and proved to be highly successful. Racing
cars even have carbon parts in their engines now, and, yes, you
guessed it, we've designed epoxies that can be used in that high-
temperature environment. After racing cars, carbon fiber composites were applied to prosthetics, a great innovation because they
are stiffer and stronger than many metals, and a great deal lighter
too. The "blades" you see being used by runners with disabilities
are made of carbon fiber composites. The material has also been
used to make bicycles, and to this day the highest-performing
bikes in the world are made of carbon fiber composites glued together using epoxies. And of course, the latest commercial passenger aircraft from Boeing and Airbus are made from carbon fiber
composites — including the one that was currently taking me on
this transatlantic journey.

Just as bolts and rivets have given way to glues and adhesives
in prosthetics and aerospace, it feels highly likely that stitches and
screws will give way to glues in hospitals. When I cut my head
open playing soccer recently, I went to the emergency room with
a bloody handkerchief pressed to my head and sat for two hours
in the waiting room. I was finally called in to be seen by a doctor,
who cleaned the wound and then produced a tube of cyanoacrylate adhesive. He squirted this on both sides of my wound, held
the sides together for ten seconds, and sent me home. This wasn't
just some crank doctor trying to save time; this treatment has become standard practice in hospitals.

Cyanoacrylate glue is best known as superglue, and it is a very
odd liquid. On its own, the liquid is an oil, and behaves like one.
But if exposed to water, the H_2O molecules react with the cyanoacrylate. This opens up the double bond that holds it together,
making it available to react with another cyanoacrylate molecule.
This creates a double molecule with an extra chemical bond ready

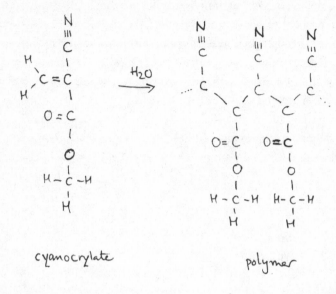

cyanocrylate   polymer

A water molecule opening up a cyanoacrylate molecule to create a polymer glue.

to react with something else. And so it does, reacting with still another cyanoacrylate molecule, creating a triple molecule with another extra bond, and then that reacts too, and so on and so on. As this chain reaction continues, a longer and more interlinked molecule is produced. This is already clever, but it gets even more so when you realize that a thin layer of cyanoacrylate liquid needs only the water vapor that's already in air to be transformed into a solid. While many glues don't adhere in a wet environment because all the water makes it impossible for them to stick to a surface, superglue works anywhere. Conversely, as anyone who's had a run-in with it knows, it makes it stupidly easy to glue your fingers together, which is why chemists are on the hunt for a way to unstick superglue quickly and comfortably.

Fingers aside, glues are holding together a lot of the world these days, and it's quite likely there's more of this ahead because, as the aircraft I was sitting in was amply demonstrating by withstanding turbulence at five hundred miles per hour, glues are up to it. We probably haven't even scratched the surface of what glues can do, especially when you consider just how many strong, sticky substances other living organisms are using. Hardly a day goes by without some scientist discovering a new glue used by plants, shellfish, or spiders.

I pondered this as I flicked through the movies available on the in-flight entertainment and hesitated when I saw *Spider-Man*. Yes, *stickiness really is a superpower,* I thought, and pressed play.

5 / FANTASTIC

I LOWERED THE WINDOW shade, shutting out the bright sunshine. This always seems such a perverse thing to do; barely a day goes by in my normal life when I don't dream of jumping through the gray clouds that perpetually hang over London to bask in the sunshine above. But having been in the sky for a while, I wanted to watch a movie and I needed it to be dark so I could see the screen properly. My neighbor, Susan, looked up sharply when I closed the shade: it affected her too. So I raised the shade a tiny bit, letting some bright shafts of light back in, and made the thumbs-up gesture, asking if it was OK to put the shade down; she nodded her consent, clicked on her overhead light, and buried herself back in her book. I felt like I'd annoyed her.

If only a screen could be more like a painting, I thought, a painting made of pigments that could change, allowing characters on the canvas to move as they would in a film. Then I wouldn't have to lower the shade at all. But no sooner had this thought entered my head than I realized that the very book Susan was reading, *The Picture of Dorian Gray*, was about a painting exactly like that. This was a bit uncanny, and in keeping with the book's spooky plot. Oscar Wilde wrote the novel in 1890, just as liquid crystals were first being discovered; he couldn't have known that they would give rise to the flat-screen technology I was using to watch *Spider-Man* — nor could he have known that liquid crystals were the very technology capable of creating the magical but sinister painting at the heart of his novel.

In the book, the eponymous Dorian Gray, a good-looking, rich young man, has his portrait painted. When Dorian sees the finished work, he's stung by the thought that he will age and become less beautiful, but the painting will not. He complains:

> It will never be older than this particular day of June . . . If it were only the other way! If it were I who was to be always young, and the picture that was to grow old! For that — for that — I would give everything! Yes, there is nothing in the whole world I would not give! I would give my soul for that!

Dorian's wish is mysteriously granted. He pursues a hedonistic life, in love with his own beauty, youth, and the sensual pleasures

An illustration of the moment when Dorian Gray first sees his youthful portrait.

they bring him, destroying the lives of others as he does so. The painting, in effect, gives him superpowers, though different from those of Spider-Man, who was now leaping across my screen. Spider-Man has super strength, the ability to cling to buildings, and a "spider-sense" that allows him to detect danger. Dorian Gray's superpower is that he never ages or becomes less beautiful; it is his picture that ages in his place. My eyes flicked across to Susan — who was now in darkness, reading *The Picture of Dorian Gray* in a pool of overhead light — and it got me thinking how hard it would be to create a moving portrait.

When you dab paint on a canvas, the liquid sticks to it, and to any other layers of paint you've already put down — just as our early ancestors learned in making their cave paintings, paint is effectively a colored glue. Thus, a paint's job is to turn from a liquid into a solid and then stay in place permanently. Different paints achieve this in different ways. Watercolor paint does it by drying, releasing water into the air through evaporation and leaving only the pigments on the page. Oil paint is made of oils — usually poppy, nut, or linseed oils. It doesn't dry. Instead, it has another trick up its sleeve: it reacts with oxygen in the air. Normally, this type of reaction is to be avoided, because oxidation turns butter and cooking oils, for example, rancid and bitter-tasting. But in the case of oil paint it is an advantage. Oils are composed of long hydrocarbon-chain molecules. The oxygen grabs a carbon atom from one chain and joins it to another through a reaction, in the process opening that molecule to further reactions. In other words, oxygen acts as a hardener (just as water acts as a hardener in superglue) — and yes, this is yet another polymerization reaction.

This reaction is extremely useful because it produces a hard, waterproof finish of plastic on the canvas (oil painting could more accurately be referred to as plastic painting); it's incredibly resilient and holds up very well with age. The polymerization takes time, though, since oxygen has to diffuse through the top, hard layers of paint before it can get to the unreacted oil underneath. This is the downside of oil paint—you have to wait a long time for it to harden. But the great masters of oil painting, like Van Eyck, Vermeer, and Titian, used this to their advantage. They overlaid many thin layers of oil paint, which one by one chemically reacted with oxygen and hardened, building up a number of layers of semitransparent plastic, one on top of another, a complex encasing of many differently colored pigments.

Layering paint gradually like this allows the artist to create wonderfully nuanced work because when the light hits the canvas, it doesn't just bounce off the top layer—some of it penetrates through to the layers below, interacting with the pigments deep in the painting and rebounding as colored light. Or, alternatively, it is fully absorbed by the different layers and thus produces deep blacks. It's a sophisticated way of controlling color, luminosity, and texture, which is exactly why the Renaissance artists adopted oil paint. Analysis of Titian's painting *Resurrection* reveals nine layers of oil paint, all working to create complex visual effects. It's exactly the intricate expressiveness of oil paint that made Renaissance art so sensual and passionate. The effect of layering is so powerful that it has transcended its roots in painting with oils and is now incorporated into all professional digital illustration tools. If you use Photoshop, or Illustrator, or any other computer graphics tools, you'll be making images in layers.

Linseed oil is also used for many applications beyond oil paint, such as treating wood, to create a transparent, protective plastic barrier—just as oil paint does, but this time, without color. Cricket bats are just one of many wooden objects traditionally given an outer coating using linseed oil. You can also go whole hog and use linseed oil to make a solid material called linoleum, again through a polymerization reaction. Linoleum, a plastic, has been used by designers and interior decorators as a waterproof floor covering. Artists use linoleum too. They carve images into it just as they do with woodcuts, to create prints. Here again, layers are the primary way of building up complexity in the final work.

A lino print, *The Secret Lemonade Drinker,* by Ruby Wright.

As visually absorbing as they are, neither printmaking nor oil painting will give you a moving image. But if you take a carbon-based molecule, one not so different from those found in linseed oil, such as 4-cyano-4'-pentylbiphenyl, suddenly a moving image becomes possible.

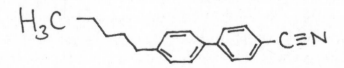

The molecular structure of 4-cyano-4'-pentylbiphenyl, commonly used in liquid crystals.

The main body of a 4-cyano-4'-pentylbiphenyl molecule is made up of two hexagonal rings. This gives it a rigid structure, but the electrons binding it together are not evenly distributed: it is a polar molecule. There are areas concentrated in negative electrical charge and some concentrated in positive electrical charge. The positive charges on one molecule attract the negative charges on another, increasing the tendency of molecules to align with one another in an organized structure—a crystal. But the tail of 4-cyano-4'-pentylbiphenyl has a CH_3 group on it, one that's flexible and wriggles, acting in opposition to the formation of a crystal. Hence, 4-cyano-4'-pentylbiphenyl structures are partly organized and partly fluid—a so-called liquid crystal.

Above a temperature of 95°F, the influence of the CH_3 tail wins, and 4-cyano-4'-pentylbiphenyl behaves like a normal transparent oil. But cool it down to room temperature, and the liquid becomes milky in appearance. It's not solid at this temperature, but something odd has happened to it. The molecules have begun to align with one another in much the same way that fish align when they're part of a shoal. It's very unusual for liquids to have a structure like this. One of the definitive qualities of a liquid is that its atoms and molecules have too much energy to stay in one place for any length of time, and so they are constantly rotating, vibrating, and migrating. But liquid crystals are different—the molecules are still dynamic, and can flow, but they keep aligned in their orientation, which has been compared to the regular alignment of atoms in a crystal—hence the name.

The alignment isn't perfect, though; because the molecules are in a liquid state, they keep moving around, swapping places, and joining other shoals. But the polar molecules give the liquid crystal another useful property—they respond to applied electric fields. They do so by changing their direction of alignment. Thus, you can get a whole shoal to point in a particular direction by applying a voltage. This turns out to be key to the technological suc-

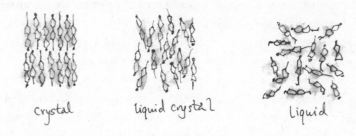

Crystal   liquid crystal   Liquid

An illustration of the difference in structure among a crystal, a liquid crystal, and a liquid.

cess of liquid crystals; it's what allows them to be integrated into electronic devices.

When light travels through a liquid crystal, subtle changes occur; the polarization changes. To make sense of this, think about light as a wave: a wave of oscillating electric and magnetic fields. But which direction do they oscillate in? Up and down, side to side, or left and right? Standard light from the sun oscillates in all of these directions. But if it bounces off a smooth surface, the surface will encourage the oscillations to move in certain directions and suppress others, depending on which ones it's aligned with. Thus, the rebounding light will contain some oscillations and not others. This is called polarized light.

It's not just surfaces that do this to light. Some transparent materials will change the polarization of light too; polarized sunglasses, for instance. The lenses of polarized sunglasses let only one direction of oscillation through. This obviously reduces the intensity of the light reaching your eyes, which is why you see the world as darker. They're especially useful at the beach, not just because they shade your eyes, but because the glare coming off the surface of a smooth sea is also polarized, and the lenses are designed to block it out. Fishermen use polarized sunglasses to help them see underwater more easily, and photographers use polarized lenses for the same reason — to cut the glare.

Some spiders can detect polarized light, and I wondered if this could be part of Spider-Man's ability to react quickly to danger, his so-called "spider-sense." In the film, he'd just narrowly escaped being captured by Dr. Octopus with an uncanny, split-second decision that allowed him to evade the villain's tentacles. The special effects were amazing, and I grinned at Susan, forgetting that despite my interest in her book, she may not have a reciprocal interest in *Spider-Man*.

Liquid crystals change the polarization of light — that's how the image of Spider-Man was being conjured up in the screen in front of me. If you put a lens from your polarized sunglasses on the surface of a liquid crystal, the light coming out of the liquid crystal will appear bright if its polarization is aligned with the lens, but otherwise it will appear dark. But here is the neat trick: if you switch the structure of the liquid crystal using an electric field, the polarization of the liquid crystal also changes. So, at the flick of a switch, you can turn the light on and off. Suddenly, you have a device that is capable of giving out white light, and then none, and then going back to white again, as fast as you can electronically switch the liquid-crystal structure — the makings of a black-and-white screen.

It may sound simple, but it took decades to achieve. It was an Austrian botanist by the name of Friedrich Reinitzer who first categorized the weird behavior of liquid crystals in 1888, just two years before Oscar Wilde wrote *The Picture of Dorian Gray*. While many scientists studied them over the course of the next eighty years, no one could really find a use for them. It wasn't until 1972, when the Hamilton Watch Company launched the first digital watch, called the Pulsar Time Computer, that liquid crystals found their moment. The watch looked great, unlike any other watch that had come before it, and it cost more than the average car. People who bought it thought they were buying the future — and they were right. Digital technology was coming, and this was the first mass-market entry in what would become a trillion-dollar industry.

The Pulsar Time Computer was made with light-emitting diodes — LEDs — which were themselves made from semiconductor crystals that emit red light in response to an electric current. They looked great, especially on a black background, and the rich and famous went crazy for them — even James Bond wore one in the 1973 film *Live and Let Die*. The drawback of LEDs at that time, though, was their high energy consumption; the batteries in those first digital watches had very short lives. In order to satisfy the sensational demand for the new digital watches, a more energy-efficient screen technology was needed. Suddenly, after decades as a lab curiosity, liquid crystals had their moment. They quickly dominated the digital-watch market because the electric power required to switch a liquid-crystal pixel from white to black is absolutely minuscule. They were cheap too — so cheap that manufacturers started making the whole display screen out of liquid crystal — this is the gray screen you see on a digital watch. The watch electronically switches certain areas of the gray liquid crystal to block polarized light, which creates black. This allows the watch to show changing numbers, so you can view the time, or date, or anything else that can be conveyed in this small digital format.

One of my strongest memories from childhood is the insane jealousy I felt when my friend Merul Patel came back to school after the holidays with the new Casio digital watch and calculator. I was ridiculously impressed as he nonchalantly pressed the tiny little buttons, which beeped happily at him. Of course, I see now it's kind of dumb — who really wants a tiny calculator? But still, at the time, I was completely captivated. It was the beginning of my addiction to gadgets.

Even though digital watches have lost their cool, they've been replaced by a seemingly never-ending parade of other digital devices, not least of which is the cellphone, which still uses liquid-

A Casio calculator watch.

crystal displays. It may seem surprising, but the same basic technology used in a digital watch also yielded the modern smart-phone screen, capable of displaying color video. This brings us right back to oil paintings, and the puzzle of creating the moving painting depicted in the novel *The Picture of Dorian Gray*. Liquid crystals could perhaps be just what's needed—but how do they create color?

We all know that if you take yellow paint and mix it with blue, our eyes interpret that mixture as green. Similarly, if you take red paint and add blue, you get purple. Color theory shows that you can make any particular color by mixing together combinations of primary colors. In the printing industry, cyan (C), magenta (M), and yellow (Y) are generally used, with the addition of black (K) liquid to control contrast. This is also how inkjet printers work and why you see the abbreviation CMYK on printer cartridges. These colors are printed onto the page, dot by dot, and it's our eyes and visual system that integrate them. We've known for a long time that the eye can be fooled in this way; Newton took note of this manipulation in the seventeenth century. In the nineteenth century it formed the basis of a painting technique—pointillism. Blobs of pigments placed close together remained physically un-mixed, so their brightness and luminosity could be controlled to create the effect the artist wanted. As predicted by color theory, it's possible to make any color by mixing paints this way, as long as the dots are small and placed close together. But changing a particular color once you've made the dots is another story. You'd have to physically change the ratios of the pigments on the can-vas. Which means you would have to remove some dots and add others. Unless, of course, you found a way to put down dots with every possible combination of colors.

This is essentially how liquid-crystal color displays work, whether they're on your phone, your TV, or, as in my case, encased in the back of the seat in front of me on the plane. We call the dots pixels. Each pixel has three colored filters that let three primary colors through. For displays these are red (R), green (G), and blue (B)—hence the abbreviation RGB. If they are all emitted equally, then the pixel appears white, even though it's made up of three separate colors. You can see this for yourself if you put a small drop of water on your phone and look through it onto the screen. The water behaves as a magnifying glass, which allows you to see the sets of red, green, and blue.

Just as the masters of oil painting had to work out how to bring darkness and shadow into their work by mixing colors and invent-ing a color theory for perception, so are today's liquid-crystal-display engineers and scientists pushing the boundaries of color display with moving images. And just as in the Renaissance, when oil painting battled it out with other techniques and mediums, such as fresco and egg tempera, so these days do liquid-crystal displays (LCDs) compete with organic light-emitting diodes (OLEDs). This battle, which is currently being played out in every new generation of TVs, tablets, and smartphones, has its own arcane language. LCDs, you might be told by an online blog, can't show deep blacks because the polarizers that keep light from coming through during a dark scene in a movie aren't 100 percent effective; you end up with grays. Similarly, because of the way color is created in LCDs, the absolute brightness of some hues suffers. Hence the issue with the window shade in the airplane cabin, and not wanting to have sunlight reflecting off the screen.

Nevertheless, the displays have got better and better, thanks to great innovations that ultimately aren't so different from layering oil paint. For example, the addition of an active-matrix layer allows some of the pixels to be switched independently from others. Thus, some parts of the image can be given higher contrast than others—instead of having to set the contrast for the whole image. This is useful for scenes of a movie that are partially lit. It is all done automatically, of course, with transistor technology—that's what the "active" means in "active matrix." Engineers have also learned to improve the way the image changes, so that you can see it well from almost any viewing angle. It used to be that the screen couldn't be seen well from certain angles, but now a "diffuser layer" has been incorporated, which spreads the light out as it leaves the screen. In comparison, the technology of OLEDs, which are the successors of the red-light-emitting diodes of the original digital watch, the Pulsar Time Computer, are now energy-efficient. They also have a much larger palette of colors and near-perfect viewing angles. But despite being much more expensive than LCDs, they're still not yet as bright.

LCDs may not be perfect, but they are essentially the dynamic canvas that Oscar Wilde dreamed of. It's now possible to have a portrait of yourself on display in your hall (or your attic) that updates daily. When liquid-crystal displays became really cheap a few years ago, people started giving them to one another as presents, in the form of dynamic photo frames. But these didn't become that popular. In fact, people hated them, just as Dorian Gray loathed his dynamic portrait. I'm convinced it wasn't the quality of the image that repelled them—plenty of people love looking at themselves on their liquid-crystal smartphone display—but rather something about the very nature of these displays. They're impostors, something fluid, magical, and dreamlike pretending to be a solid, dependable, real photograph of a moment frozen in time.

When applied to television in the form of flat-panel TVs, that same technology has been hugely popular. Switching the color of the pixels in a coordinated manner allows TV screens to display moving pictures. They're why we can see actors talking, gesturing, and making different facial expressions, and, in the case of the movie I was watching, leaping from building to building, saving the world from evil. Even though I knew what I was watching wasn't real, that it was just a collection of primary-color pixels flashing along to an accompanying soundtrack, it still stimulated me, both intellectually and emotionally, completely drawing me into the story. But here is the thing I find really difficult to understand. If I compare the experience of watching this film on a plane with standing in an art gallery viewing a masterpiece, such as Titian's *Resurrection,* I know which one is more likely to move me. It's the film, I'm afraid. I'm not proud of this. I know that Titian's paintings are great art and superhero movies played on a ten-inch display are not. Why am I so shallow? Could it be that at forty thousand feet I lose all taste in art? Or is it something to do with the heightened emotional state experienced while flying?

Static images like paintings and photographs allow us to contemplate ourselves, and how much we've changed from viewing to viewing. As we revisit great works of art, be they by Titian, Van Gogh, or Frida Kahlo, over our lifetime we can trace our reactions to them. The images may remain the same, but our sense of what they mean changes as we change. The magical liquid screens on airplanes act in the opposite manner; they are dynamic and offer us a vivid window into another world. They let us escape ourselves. Flying above the clouds, in a darkened cabin, we enter a fantasy. We can act, for a little while, like gods, looking down on the deeds of humans through our liquid portals, observing them, laughing at their foolishness, shaking our heads at their crazy ways. In doing so, our emotions are heightened. Some academic research suggests that this is due to the extreme contrast between experiencing a feeling of warm intimacy toward the characters depicted in the film, and the stark reality of sitting next to strangers in a tube while flying at four thousand feet in the air. This certainly rings true for me. I only ever cry when watching films on planes; small tears well up at even the most schmaltzy of movies, and I laugh uproariously at comedies that, on the ground, would hardly raise a smile.

By the time my film had ended, Spider-Man was victorious, but there was no record of any of the scenes I'd watched in the actual liquid crystals. They'd gone blank, ready to take on another dream. I felt less godlike. I looked over at Susan, who'd wrapped herself in a blanket and was sleeping, curled up in a position that looked comfortable, although I knew from experience that it wasn't. I was tempted to open the window shade and feast my eyes on the sunny blue skies again, but I didn't want to risk waking Susan. I wondered if I was sleepy at all, and thought I'd have a try at nodding off. I removed my shoes, reclined my seat, and tried to forget how hard it usually is for me to fall asleep on an airplane.

6 / VISCERAL

I AWOKE ABRUPTLY WHEN Susan pushed me roughly away from her shoulder, where my head had been resting. The embarrassment I felt intensified when I saw that a thin line of drool had escaped my mouth and was hanging on to Susan's sleeve. My hand came up sharply to gather it up, but I couldn't look Susan in the face to apologize, so instead I pretended I was still asleep. I slumped my head over to the other side of my seat and tried to nuzzle it into the gap between the hard polypropylene cabin wall and the acrylic seat cover. It was uncomfortable, awkward, and a little painful, but I felt I deserved this punishment. I was wide awake now, with my eyes screwed shut. How long would I have to do this before we could both legitimately pretend to forget what had happened? Was this the most embarrassing thing that had ever happened to me? No. But it was definitely up there with wetting myself at school, being explosively sick while running through a packed restaurant on the way to the toilet, and watching my grandfather sneeze over my soup just as it had been served. I replay these horror scenes from my life at regular intervals; their intensity never seems to dim. Why is it that bodily fluids are so charged with emotion? Even the term "bodily fluids" makes me uncomfortable. So many of our manners and customs are concerned with keeping our bodies' excretions in check. And yet without them we'd be in serious trouble. They are essential to our well-being when they're still inside our bodies — so why are they so loaded with disgust as soon as they leave?

"Sir, would you like the chicken curry or the pasta?"

The meal was being served. I swiveled around in my chair, pretending to have just woken up, exaggerating a groggy demeanor.

"Huh? Sorry, what?"

"Would you like the chicken curry or the pasta?"

"Er, the chicken curry. Thank you," I said hurriedly, twisting the toggle that secured my tray table.

I had not made eye contact with Susan since drooling on her, but I instinctively felt that the meal might draw a line under the episode: we both needed our saliva now.

A typical airline lunch.

I picked up the bread roll from the tray that had been placed in front of me and took a bite. It was soft, but a bit dry. Fortunately, the act of chewing the bread made it wet, thanks to my salivary glands, which had leapt into action, producing liquid that not only coated the bread, keeping it from sticking to the roof of my mouth, but also extracted its flavor. I tasted sweetness first, as my saliva dissolved the bread's sugars and delivered them to my sweet taste buds; then the bread's salty and savory qualities came through.

Taste buds require that a liquid medium deliver flavor molecules to them, which is exactly what saliva has evolved to do. Bread doesn't have its own juice, so you need saliva to enjoy it — indeed, to eat it at all. But your saliva doesn't just dissolve flavors; it also helps your gustatory system determine whether the food you're eating is nutritious, and it raises the alarm if the food contains pathogens or poisons. There are enzymes in your saliva that predigest the food, so your taste buds, and indeed your nasal receptors, can analyze what's in your mouth before you swallow. Amylase is one of the most important; it breaks down starch and turns it into simple sugars, which is why bread tastes sweeter the longer you chew it. The amylase continues to break down the carbohydrates long after you've swallowed, and it continues picking apart whatever small fragments remain in your mouth or got stuck between your teeth.

Saliva also controls the pH of your mouth, actively trying to keep it neutral. The pH scale describes the acidity or alkalinity of a liquid. The scale goes from 0 to 14, with 0 being the most acidic and 14 being the most alkaline. Pure water is neutral and has a pH of 7. Acidic liquids often taste sour, like lemon juice, for instance, which has a pH of 2. Most drinks are acidic, including orange juice and apple juice and even milk; they don't all taste sour because many also contain sugars, which help balance out the flavor profile (Coke-type beverages typically have a pH of 2.5, but the sugar in them makes them quite sweet).

A lot of the bacteria in your mouth feed on sugars and produce acid, which attacks the enamel on your teeth, giving you cavities. Which is why dentists are always telling you to eat less sugar. Saliva, though, constantly washes the bacteria away, restoring the pH of your mouth to neutral. Saliva also contains calcium, phosphate, and fluoride in a super-saturated state, which are deposited on the enamel of your teeth to repair them. Saliva contains proteins that coat the enamel, fending off acids; antibacterial compounds that kill bacteria; pain-killing compounds to soothe toothache; and other components that help clean and heal all the small cuts you get in your mouth while you're eating. In other words, your saliva is the original dental hygiene treatment, and for most other animals it's the only one. And it doesn't just protect your teeth and gums; it also keeps away halitosis (bad breath), which is caused by bacterial colonies growing at the back of your tongue.

The regular flow of saliva from your glands is constantly washing and cleaning out your mouth. To get a sense of just how much saliva you produce, go to the dentist. During treatment, a saliva-sucking machine will be placed in your mouth to get the saliva out of the way while work is done on your teeth. Your salivary glands don't take kindly to the interference, though, and replace your saliva almost as fast as it's sucked away. The average person produces about one quart of this extraordinary liquid per day.

Salivary glands are common to many species and have been evolving in animals for millions of years, for myriad different purposes: snakes have them, to produce venom; fly larvae have them, to produce silk; mosquitos have them, and while they're sucking your blood, they use them to inject you with chemicals that keep your blood from clotting. Some birds use saliva to glue together their nests; in fact, there are swallows, such as the black-nest swiftlet, which make their nest solely from solidified saliva—the main ingredient in bird's-nest soup, a Chinese delicacy.

Which brings us back to eating. Obviously, for humans, one of the main roles of saliva is to wet your food so it will slip and flow, allowing you to swallow it. Without that lubrication, things get very tricky indeed; this is perfectly illustrated in a cracker-eating competition. If you've never tried this, have a go at trying to eat as many crackers as you can in a minute without drinking water. For most people, dry crackers absorb so much of their saliva that, after just one, eating a second will scratch up their mouth, and they'll hardly be able to swallow the dry, crumbly mixture. But saliva isn't our only means of addressing the extreme dryness of some foods. This is why we often drink liquids with food. It is also why we spread fats, like butter, mayonnaise, oil, or margarine, onto dry foods: to act as a lubricant.

Most of us have enough saliva to eat whatever food we want, but some people suffer from "dry mouth," a condition that prevents the adequate production of saliva. Dry mouth can be caused by disease, but more often than not it's brought on as a side effect of a medication. It can be extremely debilitating, at times making it impossible for patients to eat solid foods at all. You can also get temporary dry mouth when you're experiencing stress and anxiety; if you're afraid of public speaking, you might have felt this while you were giving a speech — your salivary glands will slow their production, your throat will feel dry, and swallowing, even speaking, will become very difficult. You may notice that you're swallowing saliva as you're reading this; this is a common response, and it just highlights how your salivary system is closely linked to your nervous system.

Given the surfeit of saliva that dentists extract from patients' mouths, you'd think it could be treated, as blood is, and donated to patients suffering from dry mouth. But people don't want other people's saliva; it's a goo that actively repulses us — even sharing someone's drink, and the possibility of ingesting a tiny bit of their saliva, is disgusting to many. Disgust isn't the only problem with harvesting saliva, though. Saliva decomposes rapidly once it's outside the body, losing many of the properties that make it so vital. So, instead of saliva transfers, pharmaceutical companies make artificial saliva, composed primarily of minerals that protect against tooth decay, buffering compounds that control the mouth's pH, and lubricants that help to wet food, so you can swallow more easily. Artificial saliva comes in gels, sprays, and liquids. Once a loved one has used these products, or you've done so yourself, you truly start to value your salivary glands.

My own saliva had allowed me to eat my slightly dry dinner roll, which had whetted my appetite, so I turned my attention to the tiny bowl of salad on my tray. It had slices of tomato that looked a bit too big compared to the diced cucumber and shredded iceberg lettuce. The whole thing looked a bit dry and unappetizing. A small packet of dressing accompanied the salad. I tried to tear it open, succeeding only after a disproportionate struggle. The beige vinaigrette that I finally squeezed out of the packet was so viscous that it didn't coat the salad evenly, but rather sat in blobs on the tomato and lettuce, like little slugs. It turned my stomach a bit. A lot of food can be quite disgusting if you think about it out of context, which was exactly what I was doing at that moment.

Revulsion toward food is quite rare for me now, but it regularly occurred when I was a kid, and the vinaigrette slugs took me right back to those days. When I was young, my mother insisted that I eat whatever was put in front of me, and when I refused she would cite statistics about global starvation, telling me just how many people would kill for the food I was currently rejecting. This didn't help. It was disgust I was feeling, and disgust is instinctual. Rational arguments are not effective against it, as I would constantly remind her, but that got me nowhere. In general, disgust overrides moral argument, and I remember the retching feeling in my throat that came when I tried, or was forced, to eat food that repelled me. Much of what I found disgusting as a child was slimy, exactly like the salad dressing in front of me now: stuff that gloops and glops and slides and slithers. This property of liquids is called viscoelasticity: when liquids behave like solids for short periods of time, while still behaving like liquids for longer ones. This is why you can pick up slime and hold it between your fingers, which can't be done with normal liquids. Slime has solidity to it; you can feel it elastically resisting the pressure of your hands, and while most liquids fall apart, the whole piece of slime sticks together. But then, as you continue to hold it, the slime starts to flow and drip down your hand; this flow is what puts the "viscous" in "viscoelastic."

Hair gel behaves like this—you can pick it up in your hand, but it will also flow, albeit quite slowly. Thick shampoos and toothpaste are also viscoelastic. For whatever reason, in the context of personal grooming we don't find this quality so disgusting—perhaps because we don't eat any of these liquids.

It's the drippy, gloopy nature of slime that makes it disgusting—but why? Perhaps because it reminds us of our internal liquids, and their presence outside the body could signal a threat to our own health. Poo in its liquid state is disgusting, especially if you happen to stand in it, unsuspectingly, in bare feet, and feel it squishing and squirting through your toes. In contrast, a hard turd, especially from an animal like a sheep or a cow, is hardly worrying at all. Snot, in its slimy green form, is disgusting, and anyone who eats it really disgusts us. A child, however cute, with green, wet snot streaming from the nose is repellent to everyone but the parents—and even they don't usually love dealing with their kid's runny nose. It was this, actually, the snotlike nature of the salad dressing, that was repelling me now; I decided not to eat it.

But as gross as it may be, the viscoelasticity of saliva hints at some inner sophistication in its structure. One of the most important families of molecules in saliva is made up of mucins—large protein molecules most often secreted by mucous membranes. Mucus is a slimy coating that your body produces as a protective layer in places where you may be exposed to external foreign particles, toxins, and pathogens—namely, your nose, lungs, and eyes. It's the sticky stuff that streams out of your nose when you're exposed to smoke, or builds up in your eyes when dust flies into them. Mucus is sticky because the mucin proteins form a linear molecule that has many functional components, which stand ready to chemically bond to other things. In other words, it's sticky like resin glues are sticky.

The mucous system isn't always great, of course. Just look at what happens when you get a cold or other infection, and snot and green phlegm build up in your throat. Mucin molecules are

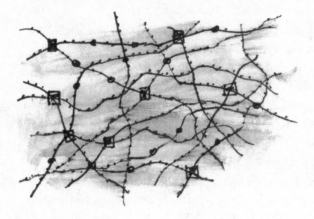

The structure of mucins, showing how their different functional components (illustrated by squares, circles, and triangles) can create a viscoelastic network that traps water to produce slimy, sticky gels.

hydrophilic, which means that they are attracted to water. They also bond to one another, creating a network of long molecules that trap water between them. This is a gel—but a viscoelastic one. The phlegm has solidity due to the mucin bonds, but because the mucin network is easily rearranged into a new structure, it flows as a liquid. In doing so, the large mucins align themselves in the direction of flow, which is why, when you drool, your saliva sticks together in long strings. Its ability to stick together, but still to flow, is what gives saliva its important lubricating qualities. Snails and slugs produce a very similar substance, which allows them to move; their mucin-loaded slime lubricates their way around the world, leaving behind little trails wherever they go. Although many people find this disgusting, as they do many slimes, snail slime is quite similar to human saliva. In fact, it is now collected and sold as a face cream. The beneficial effects of smearing snail slime on the face are as yet unproven, but this does not seem to have put off those buying it.

You might have noticed that the viscoelasticity of your saliva changes in texture throughout the day, and depending on whether you're eating, drinking, or feeling healthy or ill. Sometimes your spit is watery and very runny, and at other times it's drooly and stringy. There are, in fact, many more ways it can change in consistency, depending on which glands are producing it. Your salivary glands are controlled by your autonomic nervous system, which is responsible for regulating your unconscious actions. Salivating is one of them. There are two parts to the autonomic nervous system: the sympathetic nervous system and the parasympathetic nervous system. The parasympathetic nervous system is in charge of getting you fed properly, and produces watery saliva while you're eating. Your sympathetic nervous system takes over after eating, and helps keep your mouth lubricated and fighting infection and decay, even when you're sleeping. The saliva produced by your sympathetic nervous system has a different composition and microstructure, and as a result is thicker and stringier. It's the kind of saliva that I had unintentionally drooled onto Susan. I had a quick look at her out of the corner of my eye, without moving my head in her direction, to see if I could gauge her mood. She was eating her pasta without any obvious emotion.

I felt it was time I paid similar attention to my chicken curry. I popped a piece of it into my mouth. Something about the size of the forkful and the amount of sauce caused me to get curry on my chin. I don't know why this always happens to me, but if I'm eating sauced food and don't constantly wipe the outside of my mouth, my face gets covered. This, I am assured by others, including my nearest and dearest, is disgusting. In truth, I find it disgusting in others, so I don't know why I'm so surprised at how much it revolts other people to see me do it. It seems to be a societal norm —

having food outside your mouth is disgusting; but if the food has already been partially chewed, it's even worse. If it's mixed with saliva, or if saliva is dripping out of your mouth while you're eating, it's awful. Happily for my fellow travelers, not only do I assiduously use napkins while I'm eating, I'm not a dribbler.

Eating food is a social experience, and because the process of eating is never too far removed from feelings of disgust, table manners are incredibly important in most cultures. Babies and small children eat in a disgusting way. They lack both the coordination to get the food successfully and neatly into their mouth and the self-discipline to keep from spitting it out again, or throwing it onto the table, or the floor, or anywhere, in fact, including over their parents. One of the basic rules of our society is that we eat in an orderly manner; specifically, we don't regurgitate our food, or drool, or eat with our mouth open. So strong is the taboo associated with these behaviors that even the most savage of criminals, or the most degenerate of slobs, still generally adhere to this social norm. It's only the truly mad, deranged, or ill who defy it.

So I did my best to eat the chicken curry neatly. Soon I noticed that my forehead was getting a little sweaty. This often happens to me when I'm eating curries. The chili in the curry contains a molecule called capsaicin, which binds strongly to the receptors in the mouth that signal heat and danger. This is why eating spicy foods can produce a burning sensation in your mouth, even if the food isn't, in terms of temperature, very hot. As your mouth overheats, it's a common response for your body to try to cool you down by sweating, as I was now. Sweat is another bodily fluid that elicits disgust in others, although this is often circumstantial. If sweat starts showing through your clothes, even if you don't smell bad, you're often regarded with disgust. Having someone sitting next to you on a plane who is sweating profusely would most likely fall into this category. In contrast, sweating during sex is accepted, and in most modern societies it is thought to increase sexiness.

The University of Texas recently did a study that looked at the experience of disgust in a number of participants, using a three-domain scale: pathogen disgust, sexual disgust, and moral disgust (there being sufficient evidence that these distinct types of disgust actually do exist). To assess the level of pathogen disgust in participants, the researchers asked them questions about what their responses would be to seeing "mold on leftovers in the refrigerator, or to being presented with novel and unfamiliar foods." They assessed sexual disgust by asking the participants how they'd feel about different forms of sexual experimentation, or about having casual sex with different partners. Moral disgust was assessed by asking them how they felt about students cheating on exams to obtain better grades, or companies lying to improve profits, or other similar situations.

Ultimately, the researchers found that people who were more likely to try new and unexpected culinary experiences had a higher threshold for sexual disgust too. In fact, over all, they found that the men involved in the study had a statistically significant correlation between their mating strategies and their desire and ability to eat new and unfamiliar foods. The researchers hypothesize that men lower their disgust at particular foods in order to impress potential partners, a means of proving that they're healthy and have a strong immune system, and thus would make suitable sexual partners. In other words, eating disgusting foods could be a mating ritual, of sorts. This rings true. We know that people usually feel disgust when they see saliva, but that disgust seems to be quelled when we're sexually attracted to someone. The sloppy kiss of an aging aunt who insists on a peck on the lips—and has to clamp her hands on our face to stop us from drawing back in horror—is disgusting. But the exchange of saliva during a tongue-licking, passionate kiss with a lover is an obsessive, compulsive, wet, and visceral experience. If you felt disgust at that wetness, you'd be in real trouble from a reproductive standpoint, because lubrication during sex is important. That the same fluids that allow us to have sex are, in other circumstances, disgusting tells you something about just how much the prospect of sex reduces our resistance to bodily fluids.

All that being said, I was pretty sure that there was no way my chicken curry, or the way in which I ate it, could have been construed by Susan as some sort of mating display. I wiped the last residues of the sauce from my chin and the corners of my mouth, and then turned to the small tub of dessert on my tray. *Lemon mousse — a good choice for a palate cleanser,* I thought. But only if it's lemony enough. When our taste buds detect sourness, they stimulate our salivary glands, which then produce more saliva in an attempt to balance the pH of the mouth. This, in turn, should have the effect of washing out any lingering strong tastes, like the spices and garlic from the curry I just ate. But if the lemon mousse isn't lemony enough, I'd still be able to taste the curry while I ate it, which wouldn't be very appetizing. Happily, it had a lovely, light, foamy texture, and a strong lemon flavor, which was extremely pleasing.

Eating is more than just an exercise in gaining sustenance, more than a social ritual, and more than a mating display — it's also an emotional experience. Perhaps this has to do with the hormones that are released as we're digesting a satisfying meal, which encourage feelings of well-being, or even bliss. It's a bliss that seems to rise up from my stomach whenever I eat something good. It can even bring a tear to my eye.

We don't view tears with disgust in our society, even though they contain many of the same ingredients that make up saliva — mucins, minerals, and oils, to start. There are three sorts of tears: basal tears, reflex tears, and psychic tears. Basal tears are sort of the foundation of tears; they perform the basic function of keeping our eyes from drying out, lubricating our eyelids when we blink, and washing dust away. They also fight bacterial infection. Reflex tears wash away the many types of irritants our eyes encounter on a day-to-day basis, like smoke and dust. And psychic tears are the emotional sort, the kind you might shed after a great meal, as you're listening to sublime music, or while you're being told that your relationship is over. These tears have a chemical makeup distinct from that of basal tears or reflex tears; they contain stress hormones. The purpose of these hormones is not clear, but it most likely has to do with our desire to communicate with and get support from other people. The sight of someone crying usually elicits sympathy and the desire to comfort. Double-blind studies have shown that when men smell women's tears, it lowers their testosterone levels, and they find it less easy to become aroused.

Not that everything is about sex. But when it comes to bodily fluids, sex is never far away. Hence Susan's disgust at a stranger drooling on her.

"Are you finished with that, sir?" the flight attendant asked. He was standing by his trolley and pointed at my tray.

I handed my tray to him over Susan's lap, trying to do so in the most apologetic way possible without actually saying anything or making eye contact. It involved kind of holding the tray out to the attendant while bowing my head through my outstretched arms.

7 / REFRESHING

"TEA OR COFFEE, SIR?" the flight attendant asked, shunting his trolley along the aisle. Most of the shades were drawn down in the aircraft cabin, but the gloom was punctuated with shafts of light from a few uncovered windows, revealing an unsetting sun outside. We were six hours into an eleven-hour flight; a general feeling of lethargy prevailed. The flight attendants looked weary.

I like coffee; in fact, I love coffee. But I drink it black, as a stimulant, not for refreshment. At forty thousand feet, I didn't feel like being stimulated. On the other hand, tea made by someone who doesn't know how to make it is worse than a bad cup of coffee. *Why is that?* I thought, as the attendant looked at me with a combination of boredom and impatience.

"Tea or coffee?" he asked again.

I glanced down at my neighbor's drink, which had been placed on her tray table. It was coffee in a plastic cup with a handle so small, it wasn't actually functional. Susan had been given a plastic bag too, with packets of milk and sugar, a little stirring stick, and a napkin. It didn't look appealing; I knew I wasn't going to enjoy it. It all seemed a bit cold and institutional.

"Tea," I said, and then immediately added, "Is it hot? I mean, is it made with really hot water?" But my questions were drowned out by the drone of the aircraft engines, or maybe the flight attendant was just choosing to ignore me. He poured the tea into a cup identical to Susan's, and handed it to me on a tray with my own plastic bag of condiments.

What should a cup of tea taste like? What I'm looking for with my first sip is a savory briskness that ignites all my taste buds: not in a show-off cappuccino-with-froth-and-chocolate-sprinkles kind of way, but in a subtle, determined wave of lapping pleasure, the kind that elicits an involuntary, audible "Ah!" of satisfaction. I want to immediately taste the leafiness of the tea, not by swallowing gritty fragments of the actual leaves, but through an astringent feel in my mouth, dry enough to sweep away the taste of the stale cabin air. I want a balance to the flavor, a battle between sweet and bitter that's won by neither, with a slightly salty aftertaste. If there is sourness from the acidity, I want it to be minimal, just enough to lift the fruity, fermented flavors of the tea up to my nose and to invigorate me. Color is important; a black tea needs to be gloriously golden and transparent, and not so dark that I can't see the bottom of the cup. Ideally, I'd like to spot this before the tea has been served to me, while it's being poured out of a teapot. I also want to hear the gurgle of the liquid filling the cup, reminding me of all those moments in my life (so unlike the present one) when I've been at home with my family, drinking a cup of tea at the kitchen table.

With all that anticipation brewing, I took a sip.

It was horrible.

The tea tasted like a warm cup of flat Coke, but without the sweetness. I tasted it again to see if I'd missed anything. This time I got a twang of the unpleasant plastic taste of the cup. Out of the corner of my eye I regarded Susan, who was reading her book and sipping her coffee contentedly. Clearly, I had made the wrong choice.

But tea is reputed to be the most popular hot drink in the world. Although getting reliable facts on the topic is difficult, in Britain it is estimated that 165 million cups of tea are drunk on average every day. That compares to 70 million cups of coffee. The picture is similar in many other countries worldwide. So what does tea offer that coffee does not? And more important: why is tea often made so badly?

My cup of tea started its life as some new shoots on a seemingly unremarkable evergreen shrub that thrives only in tropical or sub-tropical climates. You could walk past this plant and never know it was the source of so much delight — our ancestors did so for thousands of years. The shrub likes humidity and rainfall, but not high temperatures, and so there are a handful of places that are ideal for growing it, like the high altitudes of Yunnan province in China, the mountains of Japan, the Himalayas of Darjeeling in India, and the central highlands of Sri Lanka. The best tea in the world, or at least the most expensive, is Da Hong Pao from the Wuyi Mountains in China; a couple of pounds can easily sell for a million dollars.

The geographical location, the altitude, and the exact conditions of the individual growing season all affect the taste of the tea leaves. One of the major headaches for tea manufacturers is to figure out how to blend tea from many different geographical locations in a way that maintains a consistent taste for their product month after month, year after year.

Although there are many types of tea, they all come from the

A tea plantation.

same plant, *Camellia sinensis.* The difference between green and black teas (and the other variants such as white, yellow, oolong, and so on) is how the leaves are processed. Every season, all the new shoots of the tea plant are picked by hand. They immediately start to wilt, which triggers enzymes that break down the molecular machinery of the leaves, turning the green chlorophyll pigment first brown, and then black. If you have ever left a bunch of herbs too long in your fridge, you will have witnessed this effect.

Green teas are produced by heating the leaves immediately after picking. The heat deactivates the enzymes, and so keeps the chlorophyll intact, and thus the green color too. Often the leaves are then rolled, which bruises their cell walls, allowing the molecules responsible for the flavor to be easily extracted. The flavor palette of green tea is made up of astringency, from a family of molecules called polyphenols (you'll remember them from the tannins in wine); bitterness, from the caffeine molecules; sweetness, from sugars; silkiness, from pectins; a savory, brothy taste from the amino acids; and a bouquet of aromatic oils. It's the careful balance of these different elements, rather than the maximum extraction of each, that yields a great cup of tea.

Black teas are produced from the same leaves as green teas — they are just prepared differently. In the case of black tea, after the leaves wilt, they are rolled, and their enzymes help break down the molecular machinery through a reaction with the oxygen in the air. This is a process called oxidation, and it changes the color from green to dark brown, producing a different set of flavor molecules. Many of the polyphenols, like the bitter tannins, are transformed into more savory and fruity-tasting molecules. Because these molecules that make up the flavor of black tea are the result of the oxidation, they are not so susceptible to being destroyed by subsequent reactions with the oxygen in the air. Thus, after drying, black teas can be stored for longer periods of time than green teas without losing their flavor.

Job done, you might be thinking. Just add water to whichever of these teas sounds best, and you'll have a refreshing drink. But tea can be ruined all too easily. Another caffeine-based drink, such as Coke, will be very similar in taste wherever and whenever you drink it. This is because the brewing process is controlled in a factory, and the flavor of the drink is not significantly impaired by being stored and transported. Thus, much of the potential for error has been removed. You can serve it at the wrong temperature (according to your preference), or in the wrong vessel (also according to your preference), but the chemical composition of the Coke is going to be reliably the same each time you order it. Inventors have long tried to do the same for tea by liquefying tea extracts to create an instant tea beverage that can be made in vending machines. So far, drinks made this way have never caught on, perhaps because they taste almost completely unlike a refreshing cup of tea. Why? The reason is thought to be that many of the key chemical components that give tea its distinctive flavor degrade and disappear soon after brewing.

The author George Orwell, most famous for classics of political fiction like 1984 and *Animal Farm,* cared so much about the problem of bad tea that he published a treatise on the drink: his eleven rules for making a perfect cup of tea. These rules include the necessity of brewing tea in a teapot, the importance of warming the pot, and the requirement that milk be added to the cup after the tea has been poured. Science doesn't offer a definitive view on what constitutes the perfect cup of tea, but does confirm some of Orwell's insights. Basically, there are four key variables that can drastically alter the quality of a cup of tea: the tea leaves, the water, the temperature of the brew, and the duration of the brewing process.

The more flavorful the tea leaves are, the more flavorful the cup of tea. But there is a catch. If we agree, even if George Orwell does not, that the best tea is the tea you personally enjoy the most, then if your favorite tea is brewed from standard teabags, it's safe to say that you're not going to find tea made from the extremely flavor-

Liquid instant-tea products.

ful and extremely expensive Da Hong Pao tea to be more refreshing. The notion of what is best is, ultimately, subjective — just as
it is with wine, and indeed with most things. On the other hand,
if you've never had the opportunity to drink a wide range of different teas (and there are approximately a thousand types available), there may yet be a more satisfying type of tea out there for
you. As with wine, a discerning knowledge of the flavor profiles of
tea requires real sophistication, and certain kinds can command
very high prices. But the tea industry is also prone to some of the
snobbish vices of the wine industry; the scarcity of a tea and clever
marketing don't necessarily translate into a high-quality product.

There is also such an enormous breadth of tea — from green tea, to oolong, to the yerba maté of South America, to the black teas of Sri Lanka — that discovering what you like can be time consuming. Personally, my perfect cup of tea changes throughout the day. In the morning, when I've just woken up, I like strong breakfast tea with milk — I find it comforting, alerting, but not too demanding. Later, I crave a black Earl Grey tea — the subtle combination of citrus and bergamot punch through the dreariness of a gray rainy afternoon.

I wondered what kind of tea Susan preferred, if any — or perhaps she wasn't a tea drinker. The problem with people who aren't tea drinkers is that I never know what to offer them when they visit me at home. "Would you like a cup of tea?" is the most welcoming phrase I know. It often rolls off my tongue before a visitor has even closed the door. The offer sounds trivial but its meaning is multifaceted: it means "Welcome to my home"; it means "I care about you"; it means "I have these delicious dried leaves that were harvested and processed thousands of miles away in an exotic climate; aren't I sophisticated?" — well, it used to mean that, when tea was first popularized in Britain, in the eighteenth century. Since then, making a cup of tea has become the default British welcome ceremony, more customary than kissing, shaking hands, hugging, or any of the other, admittedly more intimate, welcome rituals practiced in other countries. Hence George Orwell's insistence on using a teapot; it isn't just a brewing vessel, it's a physical manifestation of sharing at the heart of a home. The care and attention lavished on the teapot, the sounds of it being filled with hot water, its aesthetic appearance, the time spent waiting for the brew, and the assembly of the cups are all part of the ritual.

When performing the welcoming tea ceremony, you have to use good water. It sounds obvious, but it appears that even Orwell overlooked this variable, and given that tea is mostly water, it's easy to see how that ingredient would have a marked effect on flavor. The taste of water varies, depending on its source. The vast difference in taste between what flows from a natural spring and from a kitchen tap are obvious, but even from place to place, tap water can taste radically different. The mineral content, the organic content, and the presence of chlorine and other additives are the primary sources of flavor and smell in a glass of water. If you want to brew a lively cup of tea, you'll need to use water with a bit of mineral in it. Pure distilled water tastes flat. Too high a mineral content won't work either; the flavor of the water overwhelms the flavors in the tea, which is true of highly chlorinated water as well. Normal tap water is usually fine, but the pH of the water has to be neutral. Acidic water often has a metallic taste from the corrosion of the metal pipes that carry water from source to tap, while alkaline water often tastes soapy. Mustiness tends to come from the byproducts of microorganisms. Sometimes, especially in the morning, water has been sitting in the pipes for a long time; if the pipes are old, or made from certain metals, or if there is acidity, they can corrode a bit, giving the water an off taste. If it seems that this is happening, just let the water run for a while before you fill the kettle. If you live in an area where the water is hard — meaning it has a lot of calcium dissolved in it, usually because of the underlying geology of the region — the calcium ions inside the water will combine with the organic molecules in the tea and form a solid film that floats at the top of the cup. This is called scum. Tea scum makes tea look less delightful; it can really ruin the welcoming tea ceremony. If you have hard water, you can get rid of the scum by filtering the water or by a using a teapot that captures it on its inside walls.

Once you've secured the right water, you've got to boil it. The temperature of the brew determines which flavor molecules will be dissolved into the water, and so determines the tea's balance of flavor and color. If the temperature is too low, many of the flavor molecules won't dissolve and the tea will not only be bland, but also have a weak color. But too high a temperature can be just as bad; too many of the tannins and polyphenols that give the tea its bitterness and astringency will dissolve. Green teas have an especially high concentration of these, so they're best brewed at 160°F to 180°F, if you want to avoid an excessively bitter or clawingly astringent cup.

Caffeine is a very bitter molecule and doesn't dissolve easily in water. If you want a high-caffeine tea, then you should boil your water to higher temperatures, so more of it will dissolve in the brew. Fortunately, because black teas have been oxidized, they have a reduced number of tannins and polyphenols, which allows them to be brewed at higher temperatures without becoming extremely bitter, so you can have a highly caffeinated cup that doesn't make you wince. Black tea brewed for five minutes at 212°F will develop a dark, strong flavor, with a typical caffeine content of 0.001 ounce per cup (compared to 0.003 ounce for normal coffee). This, though, is where brewing tea onboard an aircraft can become problematic. At forty thousand feet, the pressure inside the cabin is lower than the atmospheric pressure at sea level, which lowers the boiling point of water, affecting the flavor of the brew. It's not just the initial temperature of the water that's important for brewing tea. In order for the molecules responsible for taste and color to dissolve successfully into the water, the leaves need to be in contact with the water for a specific duration of time. If the temperature of the water drops significantly during the brewing process, then fewer flavor molecules will be extracted. This will happen if you brew tea in a cold place, or if your brewing vessel is cold before you start steeping the tea, causing the hot water to drop in temperature as it warms the teapot. Hence, George Orwell's insistence that you warm the pot before making the tea. You can compensate for lower temperatures by brewing the tea for longer, but you won't get quite the same ratios of salty, sweet, bitter, sour, and savory, and the thousands of individual volatiles that provide the complexity in a perfectly brewed cup.

Here's the thing about tea: because it's so complex and many variables can affect its flavor profile (the kind of tea, the water, the brew time, and the temperature of the water), it's quite easy to lose focus. The result: you get a cup of tea that tastes completely unlike the cup you were hoping for. That's exactly what had happened to the cup of tea I was currently drinking. The flight attendants had done their best, compensating for the lower boiling point of the water on the plane with a longer brew time, and by making the tea in a warmed tall stainless-steel pot, which kept the temperature of the tea high throughout the brewing process. But it had taken them a while to get to me with their trolley, probably fifteen minutes or so since they brewed the tea, and all that time it had just been sitting there, getting cooler and less flavorful by the second. When they finally poured it into my little plastic cup, it had lost most of its fruity and leafy flavors; it had plenty of savory quality, but it was cold, bitter, and acidic, and the plastic cup itself had a distinct sharp flavor. All of this meant that I didn't get that refreshing, thirst-quenching experience I'd been hoping for — quite the reverse. It was borderline disgusting. I should never have ordered it.

But then I made another mistake. I thought I might be able to rescue the disappointing, boring brown liquid, transform it into something palatable, by using the contents of the little plastic bag the flight attendant had given me. I opened the cylindrical tub of milk and poured it into the cup, using the polystyrene stick to stir the mixture. The color of the tea turned from dark brown to milky ocher — a very pleasing color. I like milky tea. Cows' milk is sweet and contains a good amount of salt and fat. The fat in milk is shaped in small droplets, each about a thousandth of a millimeter in size, and they give a lot of flavor and a rich mouthfeel to the milk. When milk is poured into tea, those droplets of fat disperse, dominating the color and taste of the drink. They give it a malty, almost caramel flavor, and add a creaminess to the mouthfeel, which opposes tea's natural astringency. They also absorb a lot of the flavor molecules in the tea, reducing the fruitiness and bitterness, but making it creamier.

When to add the milk to your cup is a BIG bone of contention in Britain. There are those who advise adding it before the tea, on the grounds that the droplets of milk will be gently heated as more and more hot tea is added. This keeps the milk proteins from reaching temperatures that would transform their molecular structure, denaturing them and giving the milk a curdled, off flavor. Some people also argue that pouring the milk in first protects ceramic teacups from the thermal shock of the hot tea, thus keeping them from cracking; even if this was historically true, it's no longer an issue, as modern ceramics are much stronger. But for others, the very notion of pouring the milk in first is anathema. In their perfect cup of tea, you put the tea in first, and then the milk. George Orwell was in this camp, arguing that this allows you to add exactly the right amount of milk for your preferred level of creaminess.

You might doubt whether adding the milk before or after makes any difference to the taste — it being such a subtle distinction. But in an experiment now known as "the lady tasting tea," the statistician Ronald Fisher investigated this question rigorously, inventing new statistical methods to do so. In his randomized tasting experiments, he found that yes, people can taste the difference between tea with milk poured in before or after the tea.

The methods that Fisher devised revolutionized the mathematical discipline of statistics. It unfortunately did not revolutionize tea making in Britain, so even now if you order a cup of tea in a café, very rarely will anyone acknowledge that the sequence of milk and tea makes any difference. This drives me absolutely mad. Often, in a train station for instance, a server just plonks a teabag into a cup of hot water, immediately sloshes in some milk, then hands the cup to you, as if to say, "I've added all the ingredients, so it must be tea." "But you haven't asked me if I want the milk before or after," I sometimes say, when my inner rage boils over. Not that I actually want milk added before. I'm with George Orwell on this; I want the milk added after. But I still want to be asked. I'm pretty sure Orwell would agree with me on this — the current trends represent the nadir of the tea-making tradition in Britain. It's still the national drink, but coffee may well replace it if this continues, because, unlike tea, the quality of coffee served throughout the country has gone up over the past few decades, largely because of a single piece of engineering: the espresso machine.

The coffee my neighbor, Susan, was drinking began life in a more tropical setting than my cup of tea had. Coffee typically grows in forests, in countries like Brazil or Guatemala, with high summer temperatures and plenty of rainfall. To protect itself against being eaten by animals and insects, the coffee bush, like the tea plant, has evolved chemical defenses in the form of powerful alkaloids like caffeine, which can disrupt the metabolism of an organism. Caffeine's bitterness is a biological signal from our mouth, warning us that we are about to drink something that may be toxic — but in the case of caffeine, we ignore it; why is that? It's probably because we've grown to like the effect of caffeine, as well as that of other naturally derived alkaloids, like nicotine, morphine, and cocaine. But of all these psychoactive substances, caffeine is the most widely consumed. It stimulates the nervous system, relieving us of drowsiness, making us more alert. It is also a diuretic, which means it increases the production of urine. The upshot is that after you drink a strong coffee, you often need the toilet. At high doses, caffeine can cause insomnia and anxiousness. Like alcohol, caffeine goes straight into the bloodstream, so its effects are immediately noticeable; and, as with the other alkaloids, it's addictive. Once you start drinking it regularly, it can be incredibly hard to stop; the withdrawal symptoms can be severe, giving you headaches, making you tired, crabby, sluggish.

The coffee we drink is ground from beans, which are the seeds of the coffee bush. They contain a lot of carbohydrates, in the form of sugars, which give the seed the energy it needs to produce new shoots. The bean also contains proteins, which provide the core molecular machinery for the plant and instruct the seed through the reproductive process—the growth of a new coffee plant. Once the beans ripen, they're harvested, fermented, removed from the pulp, and dried. At this point, they're hard pale-green beans. The next step is to roast them; this is where the huge array of flavor in coffee is developed. You can roast your own coffee if you want; I have done it in the past. I bought raw coffee beans from my local coffee supplier, put them into a stainless-steel sieve, and held a hot-air gun over them for a while, constantly shaking the sieve. I was able to roast enough beans to make a cup of coffee in about five minutes. If you love coffee, you should really give it a go; you'll learn so much about the drink.

Roasting coffee using a hot-air gun.

The first thing you'll notice as you heat up the beans is their changing color. They'll turn yellow first, as the sugars inside the bean begin to caramelize. Then, as the temperature increases, the water inside the bean starts to boil, and pressure from the steam starts to build up; you'll know this is happening when you hear the beans cracking open under pressure. As you heat them still more, the molecular constituents of the bean start to fall apart, but they also react with one another. This use of heat is very different from its role in the production of tea leaves. There, heat is used mostly to stop chemical reactions; with coffee, the roasting actually starts the chemical reactions that produce most of the flavor. One of the most important reactions occurs between the bean's proteins and its carbohydrates. This is called the Maillard reaction, and it happens when the bean reaches a temperature between 320°F and 428°F. The Maillard reaction produces a vast array of flavor molecules; when it starts, you can immediately smell them — this is when your beans get that characteristic coffee aroma, as well as many of their savory qualities. It's the same chemical reaction that makes the delicious crust when you're baking bread, and the tasty, crispy outer layer on a steak when you're roasting or frying the meat. The reaction changes the color of the bean from yellow to brown and produces carbon dioxide gas, which will eventually go on to produce the crema foam that sits on top of a cup of coffee. At this point you'll hear crackling from the beans, as their inner structure ruptures, a result of the gas building up inside them, causing them to swell in size.

If you keep roasting the beans, you'll start to see them turn a very dark brown as the acid and tannins break down, mellowing the flavor profile. Then you'll hear a second crack as their interior structure becomes increasingly brittle and weak. You'll observe small amounts of oil leaking onto the surface of the beans at this point, signaling the complete disintegration of the bean's cellular structure. These oils, which make up approximately 15 percent of the bean, leave a glossiness on the surface that's characteristic of a French roast. If you keep roasting past this point, you'll get a shinier bean, but also a less tasty one; the high temperatures break down the molecules into smaller structures, which produce less flavor. You'll also lose a lot of the soluble carbohydrates, which are responsible for the syrupy mouthfeel of coffee. In general, the blacker the beans, the more generic and simplistic the flavor profile.

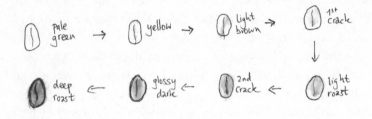

A sketch of how the color and size of coffee beans change during roasting.

When you roast your own beans, you can play around with the flavor profiles as much as you want, until you find a style that perfectly suits your palate. Doing it myself gave me a deep respect for coffee manufacturers; even with two apparently simple variables — temperature and duration of the roast — you can create an enormous range of flavors with the same beans.

Once you've roasted the beans, you've got to extract all of their flavor and get it into your cup. The earliest known methods for grinding and brewing coffee are from fifteenth-century Yemen. Arab communities there ground the coffee with a simple mortar and pestle, added it to water, and then boiled the mixture. This is still a popular way of making coffee in the Middle East; it's often called Turkish coffee. Making coffee this way gives you a very strong, dark brew; the liquid contains not just the taste compounds of the coffee, but also the grounds themselves, which affect the mouthfeel of the drink, giving it a velvety texture. But this smooth-

ness can turn gritty as you get toward the end of the cup, where
the bigger solids form a thick sediment on the bottom. Turkish
coffee is also quite bitter; brewing the grounds at boiling point allows a lot of the highly bitter-tasting molecules, like caffeine, to
dissolve into the water in large quantities. Generally, people mix
a fair amount of sugar into their coffee to offset this, resulting in a
bittersweet drink with a high caffeine content. Just what the doctor
ordered, if you want to be hopped up on a strong wallop of flavor,
combined with the one-two punch of a lot of sugar and caffeine.
But as satisfying as this can be, brewing coffee this way does eliminate a lot of the fruity flavors from the bean's fermentation, and
the nutty and chocolaty flavors you develop through the roasting.

Thus, we discover one of the biggest problems with coffee — it
often smells better than it actually tastes. Why? Because so many
of the aromas that should have been released inside your mouth
have already been released into the air while the coffee was brewing, leaving behind just the bitterness and acidity, with very few of
the aromatics. To keep from losing so much of the aroma during
the brewing process, it's best to brew at lower temperatures. This
also limits the bitterness and gives you coffee with a lower caffeine
content.

While the velvety texture of Turkish coffee can be quite pleasant,
those last sips of grit aren't great. As such, separating the coffee
grounds from the liquid became a major objective in the brewing
process — welcome the coffee filter. Making coffee by filtering it
through a fine mesh or a filter paper allows the coffee to be brewed
as the hot water comes in contact with the fine grains, but then the
liquid drips through the filter, and into another collecting vessel,
leaving the grains behind. The speed of the process is determined
by how hard it is for the water to be pulled through the grounds. If
there are too many grounds, or if the powder is too fine, then the
water takes a long time to drip through, thus dropping in temperature, which makes it impossible for the liquid to extract all of the
molecules that could give the drink flavor. Likewise, brewing the

coffee with too much water, or too coarse grains, will give you a weak cup, with little body and too much acidity, because the water won't be in contact with the grains long enough.

But if you do it right, filtering will give you a warm pot of clear, golden, grain-free coffee. There won't be a crema, though. For a lot of people, the perfect cup of coffee has a crema floating on top of the liquid—a foam created by the carbon dioxide gas that's produced during the roasting process, then released from the ground beans while the coffee's brewing. When you use a coffee filter, all the carbon dioxide is released during the filtering. No matter, though; over the past four hundred years, many other brewing methods have been invented that preserve the crema, including the French press, the Moka coffeepot, and of course, the espresso machine.

Along with producing a crema, the French press generally works faster than a filter because the coffee grounds are first mixed with the water at about 212°F, and as the coffee brews—usually for a few minutes (further brewing releases a diminishing return of flavor, and increases bitterness)—the temperature decreases to about 158°F. So, the flavor molecules are, at first, extracted quite rapidly, as the surface of the coffee powder is exposed to the hot water, but that declines as the temperature goes down, and it becomes increasingly difficult for the water to access the interior of the particles. This is when the carbon dioxide is released from the grains and escapes to the surface of the pot, trapping the liquid and forming the crema. When the coffee's done brewing, you just have to plunge the filter of the French press to stop the brewing and trap the coffee grounds. If you pour the coffee immediately, you'll have a balanced, hot cup with that pleasing crema on top. To make stronger coffee without increasing its bitterness, you can use either lots of coarse grains or fewer fine grains, the problem with the latter being that they can escape through the plunge filter and make their way into your cup, and the problem with the former being that you won't be able to extract as much flavor from the grains.

A Moka pot used to make coffee.

One way around this dilemma is to use a Moka coffeepot. In this device, the water is kept separate from the coffee grounds, in a sealed compartment. When it's heated to a boil, the water produces hot steam, which increases the pressure in the pot, eventually reaching about one and a half times the atmospheric pressure and pushing the hot water through the coffee grains and into an upper compartment once its brewed. Using the Moka extracts a lot more flavor than the French press or a coffee filter can, and it makes a strong cup. The downside of the Moka, though, is that as the water level in the boiling chamber decreases, incredibly hot steam mixes with the water, and as that steam passes through the coffee grounds, its high temperature extracts a lot of bitterness, often giving the coffee a burnt edge.

The espresso machine refines the principles of the Moka to make the most reliable — and some say the best-tasting — coffee possible. The espresso machine, so named because it can make coffee fast, in thirty seconds, heats water to between 190°F and 200°F, and then puts it under intense pressure (about nine times atmospheric pressure) before pushing it through the coffee grounds. The high pressure extracts the maximum amount of flavor and, because the system doesn't rely on steam, it doesn't overdo the bitterness and astringency. The speed of the system is important: it means there is very little time for the volatiles from the coffee to escape into the air. So you end up with a full-bodied coffee, with a great balance of nutty, earthy, and savory flavors, both fruity and acidic, with a wine-like astringency.

Because the mechanisms of an espresso machine are so controlled, it produces great coffee every time, and it's incredibly fast. This is why it's used in most commercial coffee shops, and the number of drinks you can make with it seems to have no end. Served on its own, it's called an espresso. If you add hot water to it, you've got an Americano; equal amounts of hot milk and foamed milk make it a flat white; foamed milk on its own makes a cappuccino; and so on. As with tea, milk changes the flavor profile of coffee quite radically, smoothing the astringency but also flattening the flavor profile and replacing it with a maltier, creamier taste.

Airplanes use smaller versions of espresso machines to serve first-class passengers, but the coffee served to everyone else is made with a filter. Due to the lower air pressure on an aircraft, the boiling point of water is around 200°F — which, incidentally, is perfect for coffee. That being said, coffee that's been kept warm for too long a period of time between brewing and drinking — as might happen on a plane, or in your office coffee machine — will lose a lot of its aromatic flavor, leaving you with just bitterness and astringency.

And that's not the only thing keeping you from enjoying that sort-of-hot airplane coffee. Studies have shown that our sensitivity to the five basic tastes — sweet, sour, salty, bitter, and umami — are affected by airplane noise, as well as our sense of smell. Because of this, it's impossible to taste the coffee you're drinking with the same nuance you might on the ground. This confirms my experience of flying; I generally don't enjoy coffee on planes as much as I think I'm going to.

So which drink is best — coffee or tea? Certainly, each one suits different moods and moments in life. But there are times, such as when you're on a plane in economy class, when you need to recognize that even if tea suits your mood, the chances of getting a good cup are so slim, you should just say no. I say this really as a note to myself. My cup of tea tasted terrible; the brewing temperature was too low, it was made from a bag, it had cooled down as the pot went down the aisle, the cup it was served in tasted of plastic, and the noise of the cabin dulled my senses so that whatever dismal flavor the tea had was muted. It was never going to give me that sense of contemplative stimulation I craved. In hindsight, I should have ordered coffee. Its stronger basic tastes stand up better to the cacophony of the cabin, its brewing temperature is more suited to forty thousand feet, and the filter method used on airplanes produces a balanced cup of coffee, albeit not the deepest flavor.

I noticed that Susan had finished her coffee and was about to get a refill from the flight attendant, who was walking down the aisle with a pot in his hand, raising his eyebrows at anyone who looked up. When you're in the window seat, there's never a good time to go to the loo, but I was bursting to go, perhaps because of the diuretic effects of the caffeine, and I figured if I could get past Susan before she had another cup of coffee on her tray table, I would be able to put the tea debacle behind me. I mimed my desire to exit, and she got up so that I could slide out and stagger stiff-legged up the darkened aisle to the dimly lit green signs that indicated LAVATORY.

8 / CLEANSING

AS I STUMBLED TOWARD the loo, I felt more than a little uncertain on my feet; my bad knee clicked, and I occasionally lost my balance as the airplane rumbled through the stratosphere. There were blankets half-covering slumbering passengers, and as I progressed down the aisle, I got quick snapshots of my fellow passengers' viewing habits as they watched their liquid-crystal displays: a woman singing onstage; a judge wearing a wig, looking stern in a courtroom; Spider-Man leaping. Some people were sleeping, and some were tapping away at their laptops, their faces illuminated by their screens. When I finally got to the end of the aisle, the lavatories were all occupied and I had to wait in the corridor, with the cabin crew squeezing past me to serve the passengers in business class. I glanced enviously through the gap in the curtains that separated us and caught a glimpse of reclining passengers, being ministered to as if they were Roman emperors. Then I heard the click of the lock being slid back, and saw a bright light burst forth from the lavatory. A man quickly and impassively vacated the cubicle. Did I detect a slight look of apology on his face? I braced myself for a potentially horrible smell as I entered the stall but was relieved to find that the air smelled neutral, with a hint of synthetic lemon.

I lifted the toilet seat, had a long pee, and then pressed a button to initiate the vacuum flush mechanism. I always find this a little threatening. The sucking and roaring goes on a little too long, as if to say, "Yeah, who you looking at? I could suck you down this little hole too." I turned to the basin to wash my hands and was confronted by two bottles with pump-action dispensers. I went for the one that looked more like soap and gave the bottle a couple of pumps; it squirted a clear yellow liquid into my hand. I've never really liked liquid soap; I object to the squirting action. It always reminds me of a small pet peeing with fright into your hand the moment you pick it up.

When I was growing up, liquid soap hadn't been invented. We just had bars of soap. These were so ubiquitous that basins were manufactured with indentations specifically designed to hold the soap so it wouldn't slip onto the floor or into the sink. Now bars of soap are in the minority, and becoming ever less popular. So is this progress? Is liquid soap really so much better than bars? Or is liquid soap just a modern fad, marketed to us on false pretenses, something that will eventually disappear, like flared trousers and CDs?

It's hard to say, without first understanding the advantages and disadvantages of normal soap. Soap is a miraculous substance. You can wash yourself with the clearest, purest, hottest water, but you will not get rid of any of the oily, grimy muck that's caked onto your skin. For most of history, we weren't unduly worried about this. People smelled; they were dirty. No one cared. We had bigger problems, and no sense of why soap might be important. Which isn't to say soap didn't exist. Recipes for making soap have been found on ancient Mesopotamian clay tablets dating as far back as 2200 BCE but this material has almost certainly been around longer. The process described is similar to how we make soap today: take ashes from a wood fire, dissolve them in water, and boil the solution with melted tallow (animal fat) — and magically you have a basic soap. While the Mesopotamians didn't necessarily use soap to bathe, they did use it to clean wool before weaving it into fabric. The soap removed the lanolin, a kind of grease, from the wool fibers.

But why would you use fat to remove grease? The secret is in the ash water; the Arabic word for it is *alkali*, which literally means "from the ashes." Alkalis are the opposite of acids, but both are highly reactive and can transform other molecules. In this case the alkali transforms the fats.

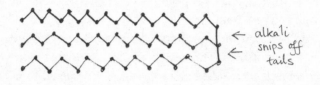

One of the main constituents of tallow, a molecule called a triglyceride. This has three tails that can be snipped off, using alkali.

Fats, such as animal tallow, are made up of carbon molecules, with a three-armed chemical structure of glycerides bound together at one end by oxygen atoms. Their structure is completely different from water's much smaller H_2O molecules. Water molecules are not just smaller than triglycerides; they are also polar, meaning the electric charges on the molecule are not distributed equally: there is a positive part and a negative part. This polarity is what makes water such a good solvent: it is electrically attracted to, and surrounds, other charged atoms and molecules, thus absorbing them. Water dissolves salt in this way, it dissolves sugar in this way, it dissolves alcohol in this way. But fat and oil molecules aren't polarized, so they can't dissolve in water. This is why oil and water don't mix.

The alkali produced from wood ash, however, splits into positive and negative components, so it dissolves in water. The resulting solution chemically reacts with the fat molecules, snipping off the three tails of the triglycerides, making them charged. This produces three soap molecules (called stearates). These hybrid molecules have an electrically charged head, which likes to dissolve in water, and a carbon tail, which likes to dissolve in oils—it's this hybrid nature that makes soap so useful.

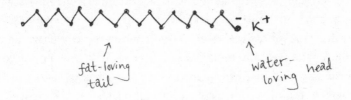

The active ingredient in soap, stearate, showing its charged head, which is "water-loving," and a carbon tail, which is "fat-loving."

When soap molecules come in contact with a blob of oil, the carbon tail of the molecule immediately buries itself in it, thanks to their chemical similarities. But the charged head of the soap wants to get as far away from the oil as possible, so it ends up just sticking out of the blob. As more soap molecules do the same, they form a molecular structure that looks like a tuft of dandelion seed: a blob of oil surrounded by a cloud of soap molecules, with their electrically charged heads sticking out.

Because the blob of oil or fat has now got a charged surface, it's become polar, and so will dissolve happily in water. This is how soap cleans — it breaks up fat and oil residue on your hands and clothes into tiny spherical blobs, which can dissolve in water and be washed away.

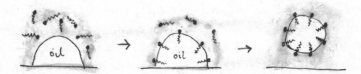

Soap cleans by the action of surfactant molecules, such as stearates. The fat-loving tail of the molecule is absorbed into oil, leaving the water-loving head sticking out. The cloud of water-loving heads surrounding the oil allows it to be dissolved in the water and so cleans a surface.

The clean, dry feeling you get from washing your hands with soap comes from the soap's ability to remove oils from your skin. At the same time, soap is slippery precisely because of its own fatty nature — it's basically modified fat. That's why it slips out of your hand so easily. It is why soaps are used as lubricants — if you are trying to remove a ring from a swollen finger, soap can slip it right off.

Using soap to clean creates a special type of liquid; it's dirty water, yes, but it's made up not just of dirt, but also of balls of fat. In effect, it's one liquid suspended in another — an emulsion. Emulsions are very useful because they allow you to suspend many different types of liquid in water. Mayonnaise, for instance, is a very concentrated suspension of oil in water, and its ratio of oil to water is roughly 3:1. You make it by shaking the two together vigorously, until they form a cream. If you did only that, though, the liquids would separate, because as we know, oil and water don't mix. But if you add a soaplike molecule, it will stabilize the droplets of oil. For mayonnaise, the binding molecule comes from eggs. Egg yolks contain a substance called lecithin, which has a molecular structure very similar to that of soap (with a fat-loving tail and a water-loving head), and when you add them to your oil-water mixture, they bind it all together and make mayonnaise. Egg yolk can also clean your hands, just as soap does, and there are plenty of shampoo recipes that use egg yolk as an essential cleaning ingredient. Mustard is another substance that can emulsify oils — which is why if you add mustard to oil and vinegar, which otherwise don't mix well, it will form a stable emulsion known as a vinaigrette. All these active substances work in the same way, and they all have a shared name: they're interface molecules called surfactants.

But soap doesn't just remove oils and fats; it also removes the bacteria attached to those oils and fats. Washing your hands with soap is the single most effective way to protect against bacterial infection and viruses. But despite the efficaciousness of soap as a cleaning agent, and its discovery so early in human development, the regular use of soap for cleanliness and personal hygiene is a modern phenomenon.

Throughout history, different cultures took very different stances on the use of soap. The Romans didn't really use it, preferring to scrape off their sweat and dirt and then bathe in first hot, then cold water, to get clean. Their public baths were an important part of their culture, and relied on a sophisticated engineering infrastructure to provide hot and cold water. In Europe, after the Roman Empire collapsed, the infrastructure that kept the public baths going fell into ruin, and so bathing went out of fashion. In crowded cities and towns, which had limited access to clean water, bathing came to be considered a health risk. During the Middle Ages, many Europeans believed that diseases were spread through miasma and bad air. They thought that washing, especially with hot water, opened up the pores and made a person more susceptible to diseases, such as the bubonic plague, also known as the Black Death. There was also a moral component associated with washing at this time; to be holy, like a hermit or saint, involved rejection of comfort and luxury. Hence, the more you smelled, the closer to God you might be perceived to be.

These attitudes to cleanliness, which seem so peculiar now, were not the norm in other parts of the world, and so visitors from the East would have found even the royal Europeans bewilderingly smelly and dirty, just as we would, looking at them from our modern vantage point. But the cultural norms of the past often seem disgusting in hindsight. It wasn't that long ago that smoking was considered quite normal, and the smell of smoke was pretty much everywhere, in offices, restaurants, bars, and trains. I still remember a time when smoking was allowed on airplanes. Now we look back with a mixture of horror and puzzlement as to how we got ourselves into that predicament. Seen in this light, the era of filthy, smelly Europeans is perhaps not so surprising.

As with smoking, the consequences of general uncleanliness were not just aesthetic. In the nineteenth century, it was still normal practice for doctors to travel from bedside to bedside without changing their clothes or washing their hands after examining women during childbirth. This practice caused incredibly high rates of maternal and infant mortality during childbirth. In 1847 a Hungarian obstetrician, Ignaz Semmelweis, mandated that physicians at the hospital in Vienna where he worked scrub their hands with a chlorinated lime solution before touching patients, and saw the death rate there fall from 20 percent to 1 percent. Despite this evidence, doctors were still reluctant to accept the idea that they might be carrying infections on their hands and transferring them to their patients, thus causing an enormous number of deaths. It wasn't until the 1850s, when a British nurse, Florence Nightingale, took up her campaign for cleanliness that this attitude was really adopted, first in military hospitals and then more widely. Crucially, she gathered statistics and invented new types of mathematical charts to show her evidence to doctors, and to the public, about the causes of disease and mortality. Gradually, as scientific evidence mounted, germ theory was accepted more generally by nurses and doctors, and hygienic washing with soaps became common practice in hospitals. Of course, not all soaps are created equal, and soap's new role in keeping people clean and healthy came at a time when industrialization and marketing were combining to create our modern Western consumerist culture. Soap was ready to transition from a commodity into a commercial product.

At forty thousand feet, in a tiny airplane lavatory, I too was hoping to use soap to make a transition: from weary to refreshed, clean, and bright-eyed traveler. I washed my hands in the tiny basin and inspected myself in the mirror. My eyes were red, and the skin around them looked dry and wrinkly; my face appeared yellow and sickly. I checked the light bulb to see if it was bluish fluorescent. It was. *Maybe that explains it,* I thought. But then, on further inspection, I noticed, with horror, that I had a speck of yellow curry sauce on my shirt collar. Susan hadn't mentioned it, but then why should she? I instinctively tried to remove it with a bit of saliva. It was just under my chin, so I had to conduct the whole operation by looking in the mirror. But the enzymes in my saliva made no headway with the yellow (probably turmeric-based) spot; in fact, wetting my collar had just made the stain spread. After five minutes of dabbing, during which there were a few rattles at the lavatory door, I had only made it worse.

Laundry powder was one of the first industrial products based on soap. Everyone has to wash clothes, and the growing importance of hygiene and cleanliness shaped attitudes toward social status and class in the nineteenth century. If you wore dirty clothes to a party, or to church or to any other religious gathering, you were deemed not just poor and of low status, but increasingly you were also held to be immoral. No longer was it true that being smelly and dirty was a signifier of virtue. Germs and disease were now associated with unclean habits. So in 1885, when the Reverend Henry Ward Beecher declared that "Cleanliness is next to Godliness," he was voicing a widely held belief that morality and spirituality had a physical manifestation, and soap was an indispensable aid to achieving this higher status.

At the same time, the spread of both railways and newspapers was bringing people together, making it possible to spread one message across a nation — soap brands were able to become national institutions. In the United States, Procter & Gamble (P&G) became the most powerful presence in the soap industry. Founded in Cincinnati in 1837 by two English immigrants, William Procter and James Gamble, the company was in the business of selling candles and soap, both made using tallow from the local meat industry. But as the nineteenth century progressed, the candle industry declined — first because of the popularity of whale oil, and then because of kerosene — while the market for soap grew. P&G invented Ivory soap and invested large sums in marketing it across the whole country, placing ads in national newspapers and magazines. Then, with the invention of the radio in the 1920s, P&G

started sponsoring serial dramas. Their audience was mostly made up of women, alone at home during the day, washing clothes and doing housecleaning; these popular dramas came to be known by a term related to the product whose manufacturer sponsored them — "soap operas."

The invention of the washing machine liberated people — primarily women — from the demanding social ritual of washing clothes, and with it came a whole new set of substances to clean our dirty laundry. Soap, the primary material used to wash clothes for almost five thousand years, suddenly got a chemical upgrade: it became detergent. Detergents are a cocktail of cleaning agents; they contain surfactants like soap, but also many other ingredients to make them more effective and less environmentally damaging. In soap, the charged, water-loving head of the molecule is attracted to the calcium in water, so if you live in a place with hard water, the calcium will attach itself to the soap and form scum, just as it does on tea. Soap scum, though, looks slightly different — it's the whitish substance you get on your hands when you wash them with a bar of soap. The scum isn't just inconvenient; it also uses up the soap, so less of the bar is ultimately available for cleaning. It can also leave an unattractive gray residue on clothes.

What to do about soapy scum? You have to make soap that's less attracted to calcium. Chemists discovered a new set of molecules similar to those of soap, with a water-loving head and a fat-loving tail; but with these new molecules they could carefully control the electric charges and make them less attracted to calcium: these were the new surfactants.

As demand for detergents grew, competition between manufacturers became intense. Companies employed the best chemists they could find in the hope of creating better ones. They developed detergents containing mild bleaches, which could better preserve whites by reacting with the molecules responsible for brown stains, and snipping them up chemically. They also put fluorescent molecules into washing powder; these molecules, called optical brighteners, attach to the fibers of white clothes and stay there after the wash. Optical brighteners absorb invisible ultraviolet light and emit blue light, giving fabric that "whiter than white" look that so many detergent companies advertise. You can see how they work if you go to a nightclub: the ultraviolet lights over the dance floor activate the fluorescent molecules in your white clothes, making them glow.

The range of surfactants expanded. Anionic surfactants (in which the water-loving head of the molecule is negatively charged, as in soap) were created not just to avoid scum formation and remove dirt, but also to keep dirt from redepositing itself on the clothes during the wash. Cationic surfactants (in which the water-loving head of the molecule is positively charged) were developed as fabric conditioners. And non-ionic surfactants (in which the water-loving head of the molecule is neutral) remove dirt even at low temperatures and are less foamy than most other surfactants. Avoiding foam is important; foam does not help to remove stains, and it is inconvenient for washing machines to fill up with foam, as it is hard to get rid of. In fact detergents often contain antifoaming agents, to suppress bubble formation.

Biological enzymes are added to most detergents in an effort to reduce the environmental impact of washing clothes. The enzymes help to chemically snip up the proteins and starches you find in stains. They're able to remove stains at lower temperatures, which makes low-temperature washing machines much more effective, thus saving energy, and money. We call the enzymes biological because they're derived from natural enzymes found in living systems that do a similar job of degrading and mopping up unwanted stuff in the body. In the UK there are two kinds of laundry detergents: bio and non-bio. The bio-detergents contain enzymes, and while these clearly wash cleaner, the non-bio detergents are still available because of a persistent myth — which has never been verified — that bio-detergents irritate skin.

We obviously care a lot about clean clothes. But we also want clean, shiny, fresh-smelling hair. Enter shampoos. The word *sham-poo* came to the English language from India, where it referred to a type of head massage using oils and lotions. The practice was imported to Britain during colonial times and eventually came to mean a type of hair washing. The first modern shampoo was created in the 1930s by Procter & Gamble; it was called Drene. Made with new, milder liquid surfactants, Drene was packaged in a glass bottle with a bright green-and-purple label. At about this same time, Unilever, Procter & Gamble's primary rival, came into business. Rivalry between these two global firms has led detergent innovation ever since.

An early advertisement for commercial shampoo.

If you look at the ingredients in a bottle of modern shampoo, you are likely to see something called sodium lauryl sulfate, or its cousin, sodium laureth sulfate. These are the building blocks of most modern shampoos; they're both very effective surfactants that do not interact strongly with calcium in water, and so do not form scum. They also do something else that we've come to consider an essential part of shampoo — they foam. And they do it very, very well.

Sodium lauryl sulfate (SLS) — notice the water-loving head and the fat-loving tail.

When you're using shampoo, foam is created while you're scrubbing, trapping air in water as you lather. The air tries to escape from the water, and when it reaches the surface of the liquid, it forms a bubble. If you're scrubbing your hair without any surfactants, the bubble will just be a thin film of pure water, which has a high surface energy with air, and so it will quickly pop. But all that changes when you add a surfactant like sodium lauryl sulfate to the mix. The surfactant's molecules will easily collect in the thin film of water that surrounds the bubble, lowering the surface energy of the liquid so much that the liquid film becomes relatively stable. As you massage the shampoo into your hair, these more resilient bubbles will keep forming, resulting in a buildup of foam. Because the surfactant is simultaneously gathering up all the oil and grease, we associate the cleaning with the foaming and judge the effectiveness of a shampoo by its foaminess. Modern ads stress this, but the foam does not help the shampoo to clean more thoroughly. Its role is purely aesthetic.

Sodium lauryl sulfate and its family of surfactants work so well and are so cheap that they've found their way into pretty much every type of cleaning product. They aren't just in shampoo, but also in cleaning liquids, laundry detergents, and even toothpastes — that's why your mouth fills with foam when you brush your teeth. Again, the role of the foam is purely for show — *Look, I'm cleaning my teeth!* it says. Because of its success, sodium lauryl sulfate ultimately supplanted bars of soap as the main way to wash the rest of the body in the shower — it heralded the advent of so-called body washes. They were distributed, just like shampoos, in small bottles and squeezy containers. And because the sodium lauryl sulfate family of surfactants is transparent, these products look great in transparent bottles, especially if you color them and scent them, as you might do for shampoo.

The appeal of body washes wasn't just aesthetic, though. When you're in the shower or the bath, bars of soap have disadvantages. As soon as they're wet, they become incredibly slippery. If you're taking a bath in an area with hard water, and you're using a bar of soap, not only do you end up sitting in scum water as the soap reacts with the calcium in the water, but you also risk losing the soap altogether if it slips from your hand into the cloudy water. And if you're in the shower when the soap slips out of your hand, it usually shoots off, pinging its way around the tub like a ricocheting bullet, potentially landing underfoot, where you might step on it, lose your balance, slip, and brain yourself. Not so with shower gel.

Shower gel also has the advantage of being contained within a bottle. Bar soap has to sit somewhere, usually on an exposed surface, where it sloughs off its outer skin of foam and slimy scum, giving it an unappealing, and certainly not very telegenic, appearance — unlike liquid soaps, which maintain their media-friendly appeal through every use. And even once it's dry again, bar soap never quite returns to its pleasing appearance straight from the wrapper; after just one use, the bar is misshapen.

In the 1980s a company called Minnetonka started thinking

about ways of bringing liquid soap out of the shower and onward to bathroom and kitchen sinks. But they knew the product needed to feel different—it couldn't be like shampoo or body wash, and certainly not like cleaning liquid—even though it would, in fact, be very similar. They had to sell it to people as new, and entirely appealing. They hit on the idea of pump dispensers, and that turned out to be a stroke of genius. Anyone who had previously worried about picking up a bar of wet soap, which had already been used by the previous person in the bathroom, could now enjoy the experience of having pristine detergent dispensed straight into the palm. The concept didn't catch on right away, though. Not everyone was impressed—for some it seemed to be an overly complicated solution to a non-problem. Others agreed with me—they didn't like the sensation that a small animal was pissing on their hand.

But if the public felt ambivalent about liquid soaps in the 1980s, the 1990s tipped the balance firmly in their favor: a bacterium called *Staphylococcus aureus*, which typically infects wounds after surgery, had, over time, developed strains that resisted antibiotics, making them very hard to treat. These strains were first discovered in the 1960s, but by the 1990s *Staphylococcus aureus* resistant to treatment by the antibiotic methicillin had become an epidemic in hospitals. In the UK, methicillin-resistant *Staphylococcus aureus* (MRSA) infections accounted for 50 percent of all hospital infections. There were similarly high rates across Europe and the United States, leading to a sharp increase in hospital mortality. By 2006, the UK had seen two thousand deaths due to MRSA, and hospitals were struggling to deal with the spread of the bacteria. Fortunately, thanks to stricter handwashing regimens—in particular, requiring nurses and doctors to wash their hands after contact with patients—the rate of death has gone down over the past decade.

Outside the hospital, though, a public-health campaign began, extolling the benefits of clean hands, and it hinged on the promotion of antibacterial soaps, which, along with sodium lauryl sulfate and its cousin molecules, contain agents like triclosan, an antimicrobial molecule. These soaps were marketed as superior to traditional soap at preventing the spread of germs. The marketing was successful —the demand for antibacterial soaps was enormous, despite the fact that no evidence proved that they were more efficacious than conventional soap and water. In fact, Dr. Janet Woodcock, director of the US Food and Drug Administration's Center for Drug Evaluation and Research, has said that certain antimicrobial soaps may not actually provide any health benefits at all.

"Consumers may think antibacterial washes are more effective at preventing the spread of germs, but we have no scientific evidence that they are any better than plain soap and water," she said in an official statement. "In fact, some data suggests that antibacterial ingredients may do more harm than good over the long term."

In 2016, antibacterial soaps were banned in the United States. But since then, liquid soaps have infiltrated everywhere. Stripped of their antibacterial agents, liquid soaps now account for the majority of soaps bought in the UK and the USA. They're still in our hospitals, our homes, and yes, our airplane lavatories, where I was now spurting some into my hand.

Bong, went the plane's intercom.

"This is the captain speaking. We're about to encounter some turbulence, so I've put on the FASTEN SEAT BELT sign. Would all passengers please return to their seats. Thank you."

Being addressed while you're in the lavatory feels a bit strange. Prior to that moment, I'd had a feeling of complete privacy, but that was shattered by a sense that the captain had just popped his head round the door. The paranoid part of my brain even considered that the announcement might just have been a ploy to get me out of the lavatory, where I'd been spending so much time reading the ingredients on the back of the liquid-soap bottle.

The liquid soap I was using did in fact contain sodium laureth sulfate. Most likely it was made from palm kernel oil or coconut oil. These trees flourish in tropical climates and have become incredibly important to the global economy because they are easy to grow and have a high yield of oil, making them stable and profitable crops for countries with a suitable climate. Fifty million tons of palm oil are produced annually, and that goes into everything from cakes to cosmetics — next time you're in a supermarket, take a look at the ingredients of cookies, cakes, chocolate, cereal, and on and on. You will very likely find palm oil in all of them.

A sketch of the structure of lauric acid, which is often obtained from palm kernel oil.

Palm kernel oil is particularly useful for making liquid soap because of its unusual chemical composition. It contains a lot of lauric acid, a twelve-carbon-chain molecule with a carboxylic acid group at the end. It looks a lot like a surfactant, but without a charged end. That's easily fixed, though, chemically speaking. Its size is what is important. Lauric acid, when used to make a surfactant, creates a hydrocarbon molecule that's much smaller than the ones found in normal soap, which are typically eighteen carbon atoms long.

Lauric acid, being smaller itself, produces a smaller surfactant and is lighter and more effective as a foaming agent. In fact, it's almost too good. Our delight with these liquid soaps has led to a huge increase in their production, and thus in the demand for palm oil and coconut oil. As a result, large sections of the rainforest in the countries where the oils are made, such as Malaysia and Indonesia, have been cut down; their immense biodiversity has been replaced with a monoculture of palm trees. This has all sorts of negative impacts, not least among them the destruction of habitat for wild animals, many of them already endangered, and the displacement of indigenous communities, who have been marginalized for centuries. Such is the demand for liquid soaps, though, and for the other products made of palm oil, that this process continues.

And to add insult to injury, detergents made with sodium laureth sulfate, which we go to such lengths to make, can actually work too well for some people. They remove fats and oils so well that they cause skin irritation, such as eczema and dermatitis. To prevent this, liquid-soap manufacturers add modifiers and moisturizers to their soaps, which replace the natural oils that the sodium laureth sulfate pulls out of your skin. You could almost be glad, then, that most of the liquid soap that's used just ends up going down the sink, without even interacting with your hands. Liquid-soap manufacturers have tried to address this by increasing the viscosity of the soap, and also by creating dispensers that squirt the soap out not as a liquid, but as a pre-formulated foam, which is more useful. The foam dispensers are actually rather good, not just because they dispense the tiny amount of surfactant you need, along with a lot of air, but also because they've finally created an active use for foam. It's not just aesthetic, as it is in shampoos, body washes, and toothpastes. In foam dispensers, the foam is the medium that carries the surfactant to your hands.

All in all, liquid soaps, of varying sorts, have become a hundred-billion-dollar industry. We rely on detergents to keep ourselves clean and scented, to keep our clothes clean and scented, to keep our hair clean and scented, to wash our dishes, and, perhaps most important in a highly populated world, to keep ourselves healthy and to stop the spread of disease — using soap is one of the most powerful ways to do this. But when we buy liquid soaps, we're mostly paying for their marketing; the essential ingredients of detergents, the ones that do the cleaning, are cheap — all the more reason to consider how these products are being made, and their impact on forests in the tropics.

Me, I love a bar of soap. It's hand-sized, and washing with it gives me the feeling of material contact, which I find reassuring and comforting. Yes, bars of soap are hard to market, but that's part of what I like about them — you buy a bar of soap because you need it, not because you think it's going to make you into a different, more successful, more desirable or sexy person.

The airplane was now rocking and bouncing around in an alarming manner. There was a sharp tap on the door, and a flight attendant asked me if I was OK. I worried for a moment that I might have been in the lavatory for hours, ranting to myself about the rise of liquid soap, but then I realized the attendant was referring to the turbulence. *Time to get back to my seat,* I thought.

But before I left the cubicle, I hesitated, my hand hovering by the second bottle by the sink. This contained another liquid, a moisturizer. Why was it there? Do we really need to moisturize our hands every time we wash them? Is this also part of the increasing pressure to consume products, regardless of whether we really need them — making soaps that clean hands too well, and then providing the antidote, a moisturizing cream? Or was I just being paranoid? I tried a squirt of it anyway; it was a nice-looking bottle, and I found the lemony fresh scent ridiculously hard to resist.

9 / COOLING

AS I WALKED BACK from the loo, I passed one of the aircraft's big oval exit doors; it had a porthole and a temptingly large red handle. I always feel a strange desire to open aircraft doors; I'm not sure why. If I did, the air inside the cabin would be sucked out, along with me and anyone else not wearing a seat belt. Everyone who was strapped in would stay put, but the air temperature on the plane would drop to approximately $-60°F$, and the air pressure would also drop, making it very difficult to breathe. At this point, as we know from the preflight safety briefing, the oxygen masks would fall from their overhead compartments.

The low air pressure at altitude is, of course, the very reason why we fly so high; the lower density of the air provides less resistance to our passage, making the aircraft more fuel-efficient and allowing it to fly farther. Nevertheless, it presents a dual problem for aircraft engineers: they have to find ways of keeping their passengers from asphyxiating and developing hypothermia. They've achieved this through air conditioning, and its history involves some of most dangerous liquids on the planet.

I returned to my seat and gave Susan an apologetic smile. I intended that one smile to convey to her that I was sorry for interrupting her reading, for making her unbuckle her seat belt, and, in forcing her to get up, for inadvertently dislodging the crumbs on her lap, though of course none of this was really my fault. It's a product of how airplane seats are arranged, and going to the lavatory is a perfectly natural thing to do anyway, even if I had been gone rather a long time.

Susan got up with a smile, which seemed to say to me: "It's fine to go to the loo, don't worry about it." She squeezed out into the aisle, and I edged past her, back into my seat. We both buckled our seat belts as the plane jolted and rocked about. The turbulence was caused by changes in the density of the air we were flying through; because of weather patterns below, we were passing through a mixture of low- and high-density air. As the plane hit pockets of high-density air, it slowed, because of the increased drag on the airplane. Then when it came upon the low-density pockets, it would drop suddenly, as lower-density air provides less lift to the wings.

But despite the rapid changes in air pressure outside, my breathing was fairly normal; the cabin pressure, though lower than what I was used to, was not fluctuating. This was thanks to the air conditioning, a field of engineering so specialized that even Einstein became interested in it, in his day, and was awarded several patents for his innovations, although at the time he was more interested in saving lives on the ground than allowing people to breathe during long-haul flights.

The problem Einstein was trying to solve was this: In the 1920s, refrigerators, newly invented, were gaining in popularity, and iceboxes, which had been the way to keep things cool for hundreds of years, were being phased out of homes. But these early fridges were not very safe. Einstein had been shocked to read in the newspaper that a family living in Berlin, with several children, had been poisoned because the pump in their fridge leaked. At the time, refrigerators used one of three types of liquid coolant — methyl chloride, sulfur dioxide, or ammonia — and all are toxic. Manufacturers had chosen them because each has a low boiling point.

Refrigerators work by pumping liquids through a series of pipes contained within them. If the temperature in them is warmer than the boiling point of the liquids, they boil. Boiling requires input of energy to break the bonds between the molecules in the liquid (called latent heat), and this heat is taken from the air inside the fridge, cooling it down. Thus the need for low-boiling-point liquids: they need to boil at the temperature inside a fridge, around 40°F. But for a fluid to be really useful in a fridge, you need to be able to turn it back into a liquid again by compressing it, using a pump.

In order to compress a gas into a liquid, you have to remove all of the latent heat from it—essentially, the heat is squeezed out of the gas. This happens at the back of a fridge—when the compressor is running, you can hear it; it is that hum that your fridge intermittently emits. It's why the back of your fridge is hot, and also why leaving your refrigerator open won't cool your home; whatever cooling is caused by the door being open is more than compensated for by the heat produced at the back by the pump, a manifestation of the first law of thermodynamics, which states that if we make something cool by taking energy out of it, then that energy has to go somewhere—it can't just disappear. So, in this case, the energy comes out of the back of the fridge.

It may sound easy to put a pump onto a set of tubes containing a liquid and then to add a valve to allow that liquid to turn into a gas, but it presents a considerable engineering challenge. The gas is under pressure, so the molecules are in constant motion, colliding against the inside of the tubes. Wherever the tubes connect with the pump there are weak points, places where, without the right materials, the constantly clamoring molecules, expanding and escaping, give way to material failure. Which is exactly what happened with early fridge designs. In the middle of the night, the ammonia leaked out and killed whole families in their beds.

Einstein wanted to do something about this, and having been a patent lawyer, he understood the technical intricacies of mechanical and electric machines. He began working with a physicist named Leo Szilard, and they set about trying to invent a new type of fridge, one that would be safer for families to use in their homes. They wanted to get rid of external pumps altogether, along with all the connectors that came with them, and instead make a system with no moving parts, which would therefore be much less likely to fail.

From 1926 to 1933 Szilard and Einstein worked together to develop different ways of manipulating liquids into gases, and then back again, to create a working fridge. Of course, as we just discovered, a liquid evaporating into a gas cools its surroundings. But going the other way, reclaiming the liquid, had always been done with a pump that forced gas molecules back into close proximity with one another, compressing them into a liquid again. There had to be a different way. Szilard and Einstein had many ideas. They built working prototypes and filed for several patents. One design used heat to drive liquid butane around a series of tubes, where it combined with ammonia to become a gas, creating a cooling effect; the gas was then mixed with water, which absorbed the ammonia and allowed the butane to be recirculated through the pipes, continuing the refrigeration process. The second had liquid metal, initially mercury, running through a series of tubes, which they vibrated using electromagnetic forces; the oscillation of the vibrating liquid acted as a piston to compress the refrigerant from a gas into a liquid — essentially creating refrigeration by having one liquid act on another liquid, without any moving solid parts. As with their other designs, the working fluids were hermetically sealed in tubes, and so, supposedly, safer than the models in use at the time.

While there was commercial interest in their prototypes — a Swedish company, Electrolux, bought up one patent, and a German company, Citogel, developed another — time was running out for the Szilard-Einstein partnership. By then the Nazis were gaining in popularity in Germany, and it was becoming more and more difficult for Jewish people like Szilard and Einstein to live and work within the country.

Szilard moved to Britain, where he came up with an invention that would change the course of history—not by cooling things down, but by heating them up. It was the principle behind the atomic bomb: the nuclear chain reaction. Meanwhile, Einstein toured Europe while an increasingly hostile Nazi party grew in power. Both Einstein and Szilard eventually ended up in America, where they were able to continue their collaboration, but by then it was too late. Scientists in America had also been working to make refrigerators safer, but they'd approached the problem the other way round—making the working fluids safer, rather than eliminating pumps. In 1930, the chemist Thomas Midgley invented Freon liquid; it was hailed as safe and cheap and put Einstein and Szilard out of the refrigeration business. Unfortunately, it turned out that Freon wasn't safe at all, but it would be fifty years before that came to light, even though Midgley was known for creating dangerous liquids.

In the 1920s, while working at General Motors, Thomas Midgley discovered a liquid called tetraethyllead, which, when added to gasoline, made it burn more completely, thus increasing the performance of gasoline engines. Tetraethyllead worked well, but it contained lead, which is highly toxic. Midgley poisoned himself while working with it. "After about a year's work in organic lead," he wrote in January 1923, "I find that my lungs have been affected and that it is necessary to drop all work and get a large supply of fresh air." Despite the clear dangers, he pressed on. It took many years, during which some of the production workers suffered lead poisoning, hallucinations, and death, but eventually, in 1924, Midgley held a press conference, demonstrating the safety of tetraethyllead. He poured the liquid over his hands and inhaled the vapor. Once again, he suffered from lead poisoning, but it didn't stop him from putting tetraethyllead into commercial production.

Tetraethyllead was subsequently used as an additive to gasoline around the world, but from the 1970s it started to be phased out, due to cumulative evidence of its toxicity (it was not completely banned in the UK until January 1, 2000). As a result, rates of lead concentration in the blood of children dropped dramatically, for example, and the social effects were widespread. A statistically significant correlation was found between the usage rate of leaded fuel and violent crime, for instance. Such is the potency of lead as a neurodegenerative substance; scientists have even speculated that banning leaded gasoline brought about a significant increase in the IQ level of people living in urban areas.

But that was all after Midgley began working on the problem of safe refrigeration. By the late 1920s, he'd found a solution. His team focused on small hydrocarbons like butane, which had a low boiling point. The downside of these substances was that they were all highly flammable and potentially explosive, which is why they're used as fuels in cigarette lighters and camping stoves.

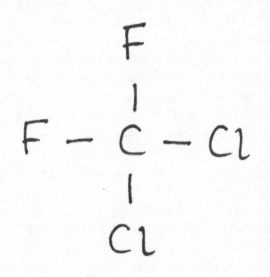

The molecular structure of the CFC Freon.

Midgley's group replaced the hydrogen atoms on the hydro-carbon molecules with fluorine and chlorine, thus creating a new family of molecules called chlorofluorocarbons (CFCs). In doing this they were potentially making something even more dangerous than the small hydrocarbons they'd started with; if these new molecules were to decompose, they would form hydrogen fluoride, an extremely corrosive and toxic substance. But Midgley's team thought that kind of decomposition was highly unlikely because the fluorine-carbon bond was so strong, the liquid would be inert. And so it proved: chlorofluorocarbons are indeed chemically inert. They seemed to be the perfect chemical solution to the problem of refrigeration because if they leaked out of the back, they wouldn't kill anyone. Midgley was right about this, but he was otherwise wrong about the safety of CFCs.

Ever since their introduction, CFCs had been leaking from the back of fridges, but it seemed the main effect of this was just that the refrigerators would malfunction — they didn't kill anyone. And because they were so cheap to produce, CFCs brought about a huge surge in the popularity of refrigerators. In 1948, just 2 percent of people in the UK owned a fridge; by the 1970s, pretty much everyone did. It was a miracle, really. We went from a nation that ran on larders and cool boxes to a place where everyone had the means to cool and store their food and drinks. It made fresh food distribution radically more efficient, cutting food waste in fish, dairy, meat, and vegetables, and so making food cheaper. It was no less than a refrigeration revolution, all thanks to the seemly innocuous CFCs.

I felt in need of a bit of refrigeration myself, sitting on the stuffy plane. I fiddled with the nozzle over my seat to try to get a bit more air. It was stuck, and I had to lift myself off my seat to get a better grip. I finally got it open, and a gale of cool air poured down on me. I must have dislodged some dust from the seat because as I sat down again, I sneezed violently. It was one of those sudden and irrepressible sneezes that you can't do anything about, but it was a serious breach of airplane etiquette, especially since I hadn't managed to catch the sneeze with my elbow. The woman in front of me turned around and peered at me through the slit between the seats, registering her disapproval. A man standing in the aisle shot me a look of unbridled hatred. My fellow passengers no doubt assumed I had the flu, or something even worse, and that I'd recklessly boarded the plane with it, no doubt ignoring my doctor's advice not to travel. This is a crime we have all been guilty of at one time or another, I suppose, and it's a fact that viruses spread fast on planes because everyone's packed so tightly into a relatively small space. I felt terrible. And, to make matters worse, the sneeze had been a little wet; it was possible that the people in the seats in front of me might have felt a droplet or two. Susan had the most reason to feel affronted, but she said nothing, apparently glued to her book. I wanted to apologize, explain to everyone that the sneeze had been caused by dust, which had probably been dislodged into the air when I'd sat down, but I didn't know how to begin. So instead I got my handkerchief out and wiped my nose and the vinyl seat cover in front of me.

Air-conditioning systems are essentially refrigerators for air. For instance, the air-conditioning system for your car's interior passes the air over copper tubes containing refrigerant, thus cooling the air. Cool air can't maintain a high concentration of water, which is why water droplets form on air conditioners (this is also why clouds form as air rises and becomes cooler). Hence, a byproduct of air conditioning is dehumidified air. In hot, humid countries air conditioning is often the only way to make traveling by car, bus, or train tolerable. But it also consumes a huge amount of energy. In Singapore, for instance, about 50 percent of the energy consumption in homes and offices is for cooling. In the United States, the entire transport sector, including trains, planes, ships, trucks, and cars, accounts for 25 percent of the country's energy use, while the heating and cooling of buildings through air conditioning accounts for nearly 40 percent.

And just as the back of your fridge gets hot as a result of cooling the interior, so too does the air conditioning of a vehicle or building release that heat back into the environment, raising outside air temperatures. The overall effect of this isn't huge except in dense cities, where the rise in temperature due to air conditioning is appreciable. Scientists at Arizona State University have shown that, solely because of air conditioning, average nighttime temperatures have increased by more than 1.8°F in urban areas. That doesn't sound like a lot, I admit, but remember, even a 3.6°F increase in global average temperatures is likely to lead to severe climate change.

Making air conditioning more energy-efficient is thus a global challenge, and one to which, I'm proud to say, I've made a small contribution. To increase the efficiency of cooling systems, the heat has to conduct quickly through the metal pipes, which is why we use copper for air-conditioning pipes. Copper may be expensive, but it's a very good conductor of heat. But on a very hot day in a stuffy office, with the outside temperatures approaching 100°F, even copper tube sometimes isn't enough to keep the room cool. The way the liquid coolant flows through the tubes can tip the balance, though.

Uniform flow, like water coming out of a pipe, is predictable, but its speed is inconsistent within the stream. Generally, the outer part of the flow, the part nearest the pipe — also called the boundary layer — is slower than the inner part. There isn't much thermal interaction between these two layers, which decreases the speed at which the heat is conducted away. The cooling system is considerably more efficient if you can achieve what's known as a turbulent flow. This is a chaotic state of flow, where the liquid tumbles and creates vortexes, mixing everything together quite thoroughly. Increasing the pressure is one way to get turbulence (turning the tap on all the way, so the water comes tumbling out of the pipe chaotically), but that uses up a lot of energy. It's better if you can disrupt the boundary layer, which we accomplish by making helical grooves inside the copper pipe so that they break up the uniform flow by constantly mixing the liquid. This has become the preferred means of getting a turbulent flow, which allows the cooling liquid to extract heat more efficiently, radically increasing the efficiency of air conditioning without any extra energy expenditure. Genius, eh?

This wasn't my invention. Einstein missed it too, though, so I don't feel too bad.

This system of creating a turbulent flow was invented in the twentieth century, at a time when I was still learning to spell and Einstein was dead. But by the time I'd gone to school, then to university, and done a Ph.D., the state of the air-conditioning sector hadn't progressed beyond it. Energy efficiency was becoming a more important issue, and there was a lot of pressure to lower the costs of making the spiral, helical-grooved copper tube. So much so, that when I finished my Ph.D. on jet engine alloys, Professor Brian Derby at Oxford University asked me to help him solve the problem. Since this problem had nothing to do with jet engine alloys, I was, understandably I think, not sure how to proceed.

Grooved copper tubes are made through a process that's pretty similar to squeezing toothpaste—just imagine that instead of toothpaste, there's a bullet inside the tube, with a diameter that's slightly greater than the nozzle, so it doesn't squirt out when you squeeze. Instead, the bullet gets pushed against the nozzle, and the tube flows around it, which stretches out the copper. But because there are helical grooves on the bullet, as you squeeze, the bullet spins and carves its grooves into the inside of the copper tube. Magic! The only problem is that the bullet had to be made by bolting together several components made from a super-hard material called tungsten carbide, and inside the massive copper-squeezing machine the pressure often got so high that the bolts snapped off, the bullet fell apart, and the whole thing ended up in a big mess and cost millions of pounds to sort out.

Miraculously, we found a liquid that solved the problem. We

determined that we could bond the two halves of the tungsten carbide bullet together by turning the inside of the material into liquid, while keeping the rest of the material solid. It's a kind of very precise welding. And as happens with a lot of discoveries, once you know the trick, it's easy to do. We just had to compress the two parts together and put them into a high-temperature furnace. This caused liquid to form *inside* the material; it flowed between the two pieces and then joined them together. Once it all cooled down, you were left with a single seamless piece of tungsten carbide. But that didn't mean the bullets would hold together through practical use. So I felt incredibly nervous when I traveled to a huge copper-pipe factory in St. Louis, Missouri, to see the first test of my tungsten-carbide bullet, knowing that if it did break, the test would cost the company tens of thousands of dollars. I am proud to say, though, that the liquid-phase bonding worked, and we filed for a European patent, *Method of liquid phase bonding* (WO1999015294 A1).

Finding ways to cool more efficiently is all well and good, but there were larger problems looming. So much work had gone into making cooling systems work better, but no one had thought about what would happen when the fridges and the air conditioners stopped working. They just went to the rubbish dump, where the valuable metals were salvaged — the steel from the frame of the fridge, and the copper tubes. No one collected the CFCs; they evaporated quickly, as soon as the copper pipes were cut, cooling them one last time as the liquid evaporated into thin air. No one was worried about them. CFCs were already being used as propellants in cans of hairspray and other disposable items: they were supposedly inert, so what harm could they do? It was just assumed that once they became a gas, they'd be dispersed by the wind. Which is exactly what happened. But over the course of decades they found their way into the stratosphere, where the ultraviolet light from the sun started to break them down into molecules that could do us a lot of harm.

The sun emits light we can see and light we can't see. Ultraviolet light is the latter. It's the light that gives us a tan, and because it has so much energy, it can and does burn us: prolonged exposure can damage your DNA, and eventually causes cancer. This is why wearing sunscreen is essential; the job of this liquid is to absorb ultraviolet light before it hits your skin. But there's another barrier between you and the ultraviolet light that's a lot more effective — the ozone layer. Ozone is like a sunscreen for the planet, and like sunscreen, it can't really be seen once it's been applied. In fact, our plane was currently flying through the ozone layer, but looking out the window, you'd have no idea.

Ozone is related to oxygen. The oxygen we breathe is a molecule made up of two oxygen atoms bonded together (O_2); ozone is a molecule made up of three oxygen atoms bonded together (O_3). It's not very stable, and being highly reactive, it doesn't stick around for long. Ozone also has a smell, which you can sometimes detect during the production of sparks — some of the O_2 in the air is transformed into O_3 as it encounters the spark's high energy, and the resulting reaction produces a strange pungent smell. But while there's not a lot of ozone in the air we breathe down on here on terra firma, up in the stratosphere there's enough ozone to form a protective layer that absorbs ultraviolet light from the sun. But when CFC molecules find their way into the ozone layer, they break down after interacting with the high-energy rays of light emitted from the sun. This creates highly reactive molecules called free radicals; these then react with the ozone and decrease its concentration, thus depleting our ozone layer.

By the 1980s, atmospheric scientists had begun to realize that the effect of CFCs on our ozone layer was significant, and had huge consequences. In 1985, scientists from the British Antarctic Survey reported that there was a hole in the ozone layer, spanning eight million square miles, above Antarctica, and not long afterward it was determined that across the globe, the thickness of the ozone layer was degrading. CFCs are, by and large, to blame for this, and so an international ban, called the Montreal Protocol, was put in place and took effect in 1989. CFCs in refrigeration were banned, as was their use in dry cleaning, where they were used instead of water to clean clothes. But despite the swift response of the global community, there are still CFCs in circulation, and other holes have opened up in the ozone layer. In 2006, a hole measuring one million square miles was found over Tibet, and in 2011 there was a record loss of ozone over the Arctic, which suggests we won't be able to recover from all this damage until the end of the twenty-first century.

But back in the CFCs' heyday, chemists spent a lot of time exploring the properties of carbon- and fluorine-based molecules. They discovered an amazing family of molecules called perfluorocarbons, or PFCs. Unlike CFCs, PFCs don't contain any chlorine —they're liquids made entirely of carbon and fluorine atoms. The simplest PFCs resemble hydrocarbons in which all of the hydrogen atoms have been replaced by fluorine atoms.

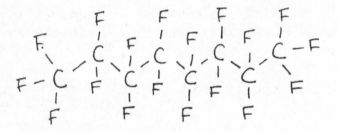

The molecular structure of a perfluorocarbon molecule.

Fluorine bonds are extremely strong, so they're also very stable, making PFCs very inert. You can dunk pretty much anything you like into them with impunity, even your phone, which will continue to operate as if nothing had happened. You could put your laptop in a bucket of PFC—and people do, because the liquid cools them down during operation much more efficiently than their internal fans do, allowing the computers to operate at much higher speeds. But even more miraculous than that is the fact that PFCs are able to absorb a high concentration of oxygen—up to 20 percent of their volume, in fact—which means they can act as artificial blood.

Blood substitutes have a long history. Blood loss is a major cause of death, and the only way to get more blood into people is through a transfusion. But for a successful transfusion, you can't use just any blood. Human blood isn't all the same type; transferring blood from one person to another is successful only if their blood type matches. A scientist named Karl Landsteiner discovered blood types in the early 1900s, and classified them as A, B, O, and AB. In 1930 he was awarded the Nobel Prize for this insight, and a decade later the enormous casualty count of the World War II led to the establishment of the world's first blood banks.

But because of the challenges of matching donated blood with patients, scientists have been on the hunt for a reliable synthetic blood, which would eliminate the need to match blood types and lessen the strain on blood banks. In 1854 some doctors used milk, with a degree of success, but it was never taken up by the medical establishment at large. Some people have also tried to use blood plasma extracted from animals, but that was found to be toxic. In 1883 a substance called Ringer's solution was developed, a solution of sodium, potassium, and calcium salts that's still used today, but as a blood-volume expander, rather than a true substitute for blood.

It wasn't until PFCs came along, though, that people really started to believe a viable artificial blood could be created. In 1966 Leland C. Clark Jr. and Frank Gollan, two medical scientists from the United States, began studying what would happen to rats if they inhaled liquid PFCs. They found that the mice were still able to breathe, even when fully submerged in a bath of liquid PFC, and then were able to breathe air again upon removal—effectively transitioning from a fishlike existence, where they obtained their oxygen from the PFC liquid, back to a mammalian one, where they got their oxygen from air. This so-called liquid-breathing appears to work not just because their lungs could obtain oxygen dissolved in the PFC, but also because the liquid can absorb all the carbon dioxide the mice were exhaling. Further studies have shown that mice can liquid-breathe for hours, and research continues, with the ultimate aim of figuring out how humans might be able to liquid-breathe. In the 1990s the first human trials were conducted. Patients with lung problems were asked to liquid-breathe, using PFCs that were loaded up with medication for their lungs. The therapy seems to work, but for the moment, not without side effects.

No one is quite sure where this strange technology might lead, but if PFCs do become prevalent in one way or another, we'll need to work out their potential environmental impact. The world has managed to avoid catastrophic loss of the ozone layer by banning CFC liquids and replacing them with fluids less damaging to the environment — these days the refrigerant in your fridge is likely to be butane. It's a highly flammable liquid, and if it leaks from the back of your fridge it could be hazardous, but it's still safer than the liquids used in Einstein's day, and it's a much better bet for the planet. Our protective sunscreen layer of ozone is too precious to destroy with CFCs.

But while the risk of using butane may be small enough for refrigerators, it's still too great for the engineers of aircraft. These days liquid refrigerants aren't used in aircraft air-conditioning systems. Instead, air is actually sucked in from outside the plane, and through a series of compression and expansion cycles is used to cool the interior — it's very cold out there, after all. The downside of this, though, is that when the plane is on the tarmac, the air conditioning doesn't work very well because the air on the ground is warmer. Which is why, adding to the general pleasures of sitting on a delayed flight, when you're stuck on a plane on the tarmac, waiting for takeoff, it can be sweltering.

A plane's air-conditioning system does more than just regulate temperature and humidity, though; it's also set to equilibrate the air pressure inside the cabin. At forty thousand feet, the air outside doesn't have enough oxygen for people to breathe easily—or at all. So the air pressure inside the cabin has to be a lot higher than the air pressure outside. This puts the skin of the fuselage in essentially the same state of stress as an inflated balloon, causing the aircraft to bulge. The bulging can lead to cracks, so to minimize the chances that they will form, the air-conditioning system makes a compromise: the pressure is set to be high enough to allow people to breathe normally, but not so high that the aircraft skin is put under undue stress. As the plane descends, the air-conditioning systems pump more air into the cabin to equilibrate to pressure levels on the ground, which is why your ears pop.

Planes don't carry liquid oxygen on board for emergencies. In case of a loss of cabin pressure, the masks that drop from the overhead compartment will supply you with oxygen made by a chemical oxygen generator—it creates oxygen gas through a chemical reaction, allowing it to be very compact and lightweight, both essential features for anything carried onboard an aircraft. I've never been on a flight during which oxygen masks have been deployed, and I'm fascinated by how well those systems are hidden. I was inspecting the overhead compartment, trying to figure out how it works, when the flight attendant leaned toward me with some urgency. He passed me a card. At first I was puzzled, but then I realized we must be approaching San Francisco. It was time to fill in my customs declaration form. For that I was going to need another liquid—ink.

10 / INDELIBLE

I FOLDED THE TRAY table down and placed the customs form on top of it. I needed a pen. Did I have one? I couldn't remember. I checked my jacket pockets. Nothing. My carry-on bag was under my feet, but I couldn't bend down far enough to search through it because of the tray table. I tried anyway, pressing my face into the tray table as I reached for my bag below. It was awkward. I knew I should have just folded the table up, but for some unaccountable reason I didn't. I'd managed to get both hands into my carry-on, and they were feeling about, exploring the unseen world of my bag. By touch I identified my phone, the adaptor for my laptop, and some socks. Because my face was turned toward Susan, I ended up grimacing at her. Her eyes flicked across at me and seemed to register exasperation, as if I were a small child looking for attention. Then I hit gold. At the bottom of my bag I came across something that felt cylindrical, like a pen. Like a pearl diver heading for the surface, I lifted my head and extracted the object from the deep recesses of my bag. It was indeed a pen, albeit a pen I had no recollection of putting in my bag, or really of owning or buying in the first place. It had stayed there, unobserved among the detritus of my life, the small change and the chocolate wrappers that accumulate over time without my ever thinking I would need it. It was a ballpoint pen.

A ballpoint is the essence of pen-ness: it doesn't have the social status of the fountain pen, nor the sophistication of the fiber pen, but it works on most paper and does the job you need it to do. It rarely leaks and ruins your clothes, and it can lie unattended at the bottom of your bag for months and still work the first time you try to use it. It does all that, and still costs so little that it's routinely given away without thought. Indeed, most people regard ballpoint pens as common property: if you give someone such a pen to sign a form and they forget to give it back to you, you wouldn't brand the person a thief; you probably wouldn't even remember where you got the pen in the first place — it's quite likely you took it from someone else. But if you think what makes ballpoint pens so successful is their simplicity, you're wrong. That couldn't be further from the truth.

Obviously, what you need in a pen is ink. Ink is a liquid designed to do two things; first, to flow onto the page, and then to turn into a solid. Flowing is not difficult; it's what liquids do. And turning into a solid is something they generally do too. But doing both in the right order, reliably, and in a pretty fast time frame, so the ink doesn't smudge and become unreadable, is much trickier than it looks.

Historians believe that the ancient Egyptians were the first people to use a pen, around 3000 BCE. They used reed pens, usually made of bamboo or another reedy plant with stiff, hollow shoots. By drying the shoot and shaping its end with a cutting tool, creating a fine tip, they made a good vehicle for ink. The shoots had to be just the right size for the pen to work, though; if the diameter of the tube was narrow enough, the surface tension between the ink and the reed surface would slow the force of gravity and hold a small quantity of ink in place. Once the reed came in contact with the papyrus, which the Egyptians used as paper, the ink would be sucked onto the papyrus fibers through capillary action — the same force responsible for wicking, in candles and oil lamps. As the dry fibers absorbed the water in the ink, the pigments would stick to the surface, and once the water evaporated completely, the ink marks would hold on to the papyrus permanently.

The Egyptians made black ink by combining soot from oil lamps with the gum from the acacia tree, which acted as a binder. Like the resin that glued together their plywood, the Egyptians used acacia tree gum as a glue to bond the black carbon of the soot to the papyrus fibers. And because carbon is hydrophobic, meaning it doesn't mix with water, the acacia gum also allowed the carbon to be incorporated into water, creating a smooth, black, free-flowing ink. Gum arabic, as it's called, is still used today; you can buy it in most art shops. The proteins in the gum allow it to bind to many different pigments, and hence, it can be used to make all kinds of coloring agents — watercolors, dyes, and inks, to name a few. But the Egyptians used carbon, and that, as it turns out, was a good choice. Carbon-based ink is easy to make and very unreactive, which is why we have Egyptian documents going back thousands of years, preserved for us by the chemical permanence of carbon black ink.

Job done, you might think. But carbon ink is not perfect. It wouldn't be good for filling in customs forms, for instance, be-

A fragment of papyrus from the *Book of the Dead of the Goldworker Amun, Sobekmose* (1479–1400 BCE).

cause, being water-based, it doesn't dry fast and hence is easy to smudge. And when it does dry, the sooty pigment isn't held strongly to the writing surface by the gum — so you can mechanically rub it away. Maybe you don't care, but others did, and so began hundreds of years of experimentation in the hope of making something better.

Eventually, they found gall ink: the ink Christians used to write the Bible, the ink Muslims used to write the Koran, the ink Shakespeare used to write his plays, the ink lawmakers used to write their Acts of Parliament. Gall ink is so good that it was commonly used right up to the twentieth century.

You make gall ink by putting an iron nail in a bottle with some vinegar; the vinegar corrodes the iron and leaves behind a redbrown solution, full of charged iron atoms. This is where the galls come in. Galls, also known as oak apples, are growths that sometimes turn up on oak trees. They're created when wasps lay their eggs in an oak bud. As the bud develops, the wasps manipulate the molecular machinery of the oak bud to create food for their larvae. This is bad for the tree, but good for literature, because it produces oak galls, with their high concentration of tannins, which led to a revolutionary innovation in ink.

Tannins are widely found in the plant world; they're part of a plant's chemical defense system, and yet somehow we've developed a taste for them — you may recall that the tannins in tea and red wine give them their astringency. Tannins are colored molecules that are good at chemically bonding to proteins — and thus they are able to impart color through bonding to things made of proteins. They have traditionally been used to stain leather, which has a high percentage of the protein collagen — hence the origin of the term "to tan leather." They're also a big part of why red wine and tea can leave such bad stains on your clothes and teeth. So, the use of tannins in ink is perhaps not so surprising, ink being, essentially, an intentional stain. But it's hard to create a liquid with a high concentration of tannins — that's where the iron-vinegar solution comes in. It reacts with the tannic acid from the galls and produces a substance called iron tannate, which is highly water-soluble and very fluid. When iron tannate comes in contact with paper fibers, it flows, through capillary action, into all the small crevices in the paper, distributing itself evenly. And as the water evaporates, the tannates are deposited inside the paper, leaving a lasting blue-black mark. Its permanence is its great advantage over carbon inks: because the pigment isn't stuck to the surface of the paper, but rather within it, it can't be removed by rubbing or washing.

Of course, the very indelibility of gall ink was also one of its drawbacks for those who wrote with it. The ballpoint pen I was currently using to fill out the customs form didn't require me to dip the pen nib into a reservoir of ink, so there wasn't any ink coating the outside of it. My fingertips were still as clean as they were when I'd washed them with the liquid soap earlier. For most of the history of writing, this was emphatically not the case. Ink would get everywhere, especially on writers' hands, and gall ink, being very permanent, did not come off easily—certainly not by washing with soap. People complained, and ironically, some of these complaints came to be written down in gall ink. By the tenth century, the caliph of the Maghreb (now the region of northwest Africa encompassing Algeria, Libya, Morocco, and Tunisia) had had enough; he demanded a solution from his engineers. In due course, in the year 974, he was presented with the first fountain pen recorded in history. This pen held its reservoir of ink within it, and apparently did not leak, even when held upside-down—I have to say, though, that this seems unlikely, not because the engineers of the time weren't ingenious, but because the fountain pen got reinvented many times over the next thousand years, and it was only after many, many iterations that a reliable fountain-pen mechanism was created, in the late nineteenth century. Leonardo da Vinci had a go in the sixteenth century, and there's some evidence that he was able to make a pen that wrote with continuous contrast, whereas ink laid down by quill pens, which were in common use at the time, tended to fade in and out. And there were certainly fountain pens around in the seventeenth century when Samuel Pepys mentioned them in his diary, content as he was at being able to carry around a pen without also needing to tote an inkpot. But those fountain pens weren't perfect; he still preferred using a feather quill pen and, yes, gall ink.

The nineteenth century saw a big surge in fountain-pen patents. But though all these pens had free-flowing ink, no one had yet devised a means of controlling the flow so that the ink didn't all rush out at once, making an enormous blob on the page. They couldn't just make the opening to the reservoir of ink very small: a tiny hole blocked the ink from coming out at all, and with a medium-sized hole it glugged its way sporadically onto the page. The reason for this behavior, which fountain-pen inventors were slowly beginning to understand, was the influence of air and the formation of vacuums inside the ink reservoirs.

When you try to pour liquid out of a container, you have to replace it with something; otherwise a vacuum will form inside the vessel, preventing more liquid from flowing out. You'll notice this if you try to drink from a bottle while covering the whole aperture with your mouth; the liquid comes out in glugs as the air fights to get in and replace the liquid you're drinking. Each glug corresponds to air forcing its way into the bottle, and as it does so, it keeps the liquid from coming out. They take it in turns—liquid out, air in, liquid out, air in, glug, glug, glug. If you leave the mouth of a bottle partially open as you drink, then you'll be able to drink continuously, without any glugs, because the air can flow in more smoothly. That's why it's easier to drink from wide-mouthed vessels, like cups and glasses.

But early fountain pens didn't have any mechanism for getting air into the reservoir of ink, so it was hard to get a consistent flow of ink onto the page. Putting a hole in the top of the reservoir seems like the obvious solution, but if you turn the pen upside-down, it will leak everywhere. The problem left everyone pretty flummoxed until 1884, when an American inventor, Lewis Waterman, perfected the design for a metal nib that allowed ink to flow down a groove by a combination of gravity and capillary action, while incoming air passed through in the opposite direction toward the reservoir. His design ushered in the golden era of the fountain pen, the cellphone of its age, transforming the way people communicated and making pens a highly coveted possession. Having a fountain pen signified you were important — you were someone who needed to be able to write anywhere, anytime. Just like the early cellphones, or the first laptops, or any number of more recent gadgets: it was cool.

But inevitably, another problem arose. Gall inks are often highly acidic, so they corroded the new metal pen nibs. They also often contained small particulate matter, which was visible in the ink when you wrote on the page, or clogged up the nib so the ink couldn't come out. People would shake their pens in rage, trying to dislodge whatever unseen obstacles were mucking up their writing, but in the process, they would lob ink across cafés or onto the clothes of unsuspecting passers-by. The fountain pen may have been perfected — the ink was not. It was time to replace gall ink.

But that was a complex problem. The particular chemistry of an ink and its ability to flow within the pen but not corrode it, its reaction with the paper, its ability to create a permanent mark but also dry quickly — all had to be considered at once. To use engineering jargon, it's a multiple optimization problem. Ultimately, there were many solutions, and each pen manufacturer incorporated a different one into its design, which is why, if you buy a fountain pen, the manufacturer will insist you use its specially formulated inks. The Parker Pen Company, for instance, developed Quink ink in 1928 to combat the problem of blotting. Parker's chemists combined synthetic dyes with alcohol to create an ink that flowed well in the pen, but then dried very quickly when it came into contact with paper. Unfortunately, it also chemically attacked some of the plastics, such as celluloid, that they'd started using to make pens. It also wasn't water-resistant, so if the paper got wet, the ink would start flowing again, often separating out the individual dyes used to make up the ink—black ink would separate into yellow and blue, for instance—ultimately making the writing unintelligible.

But despite all the problems, most pen manufacturers were convinced that fountain pens were the future, and that optimizing the ink was the answer to a reliable portable writing instrument. But the Hungarian inventor László Bíró had a completely different idea. He turned the optimization problem on its head. Before becoming an inventor, he'd worked as a journalist and had noticed that the inks used by newspaper printers were excellent—they were extremely fast-drying, and rarely smudged or formed blots. But they were too viscous for a fountain pen; they wouldn't flow, and gummed up the pen. So, he figured, instead of changing the ink, why not redesign the pen?

Bíró's newspaper articles were printed on a press made from a set of rolls that press ink onto a continuous sheet of paper. In order to have the millions of newspapers needed to meet the demand across the country ready for overnight delivery, they had to be printed very quickly. The pages went through the press at a rate of thousands per hour, so it was imperative that the inks dry immediately; otherwise they would smudge as the pages were assembled into a newspaper. To meet that need, the printing ink that Bíró admired so much was invented. As he considered how to make a better pen, he thought about ways of re-creating the printing process on this much smaller scale. He'd need some sort of roller that could continuously ink the pen tip; eventually, he hit upon the idea of using a tiny ball. But how to get the ink onto the ball so it could then be used to roll the ink onto the page? He'd assumed that the printer inks would be too thick for gravity to pull them down from the pen's reservoir to the ball. But a strange piece of physics came to his rescue—non-Newtonian flow.

There's a relationship between the speed of a liquid's flow and the sheer force that's exerted on it—what we call viscosity. So, thick liquids like honey have a high viscosity and flow slowly, while runny liquids like water have a low viscosity and flow quickly under the same force. For most liquids, if you increase the force you're applying to them, the viscosity will remain the same. This is called Newtonian flow.

But some liquids are strange; they don't play by the rules of Newtonian flow. For example, if you mix cornmeal with a bit of cold water, it forms a liquid that's runny when you stir it gently, but if you try to stir it quickly, the liquid becomes very viscous, to the point that it behaves like a solid. You can punch its surface, and it won't splash at all, but rather resist your fist, as a solid does. This is what we call non-Newtonian behavior—the liquid doesn't have one viscosity that determines its flow.

That cornmeal liquid is sometimes referred to as oobleck (the name comes from Dr. Seuss's book *Bartholomew and the Oobleck*). Oobleck's non-Newtonian behavior is entirely due to its internal structure. On a microscopic level, oobleck is full of tiny starch particles, like cornmeal suspended very densely in water. At low speeds, the starch particles have enough time to find routes to flow around one another—a bit like passengers leaving a packed train. This is when they flow normally. But when put under pressure to flow quickly, as they are when you're trying to stir the oobleck quickly or punching its surface, the starch particles don't have enough time to move around one another, and so they're stopped in place. And just as passengers at the back of a train cannot move if those at the front are stationary, so too does the thwarting of a few starch particles hold up the rest, which is why the whole liquid locks up, becoming more and more viscous.

Oobleck is not the only non-Newtonian liquid. If you've ever painted a wall with emulsion paint, you might have noticed that the paint is extremely thick when it's in the can, almost like a jelly. But if you follow the instructions on the side of the can and thoroughly mix the paint, you'll find that as you stir, the paint becomes fluid, and then turns back to jelly as soon as you stop. This is also non-Newtonian behavior, but here the liquid is becoming runnier as a result of the force exerted on it, rather than more viscous. Again, the reason originates from the inner structure of the liquid. Emulsion paint is just water with lots of tiny droplets of oil held in suspension inside it. When the droplets are allowed to settle, they're attracted to one another and form tiny bonds, trapping the water between them to form a weak structure — a jelly. When you stir the paint, the molecular bonds holding the tiny droplets of oil to one another are broken, releasing the water and allowing the paint to flow. The same thing happens when you put the paint under stress by spreading it onto a wall with a paintbrush. But once the paint is on the wall and it's no longer under stress, the bonds between the droplets of oil re-form and the paint becomes viscous again, creating a thick coat that doesn't drip. That's the theory, anyway; obviously it all comes down to how well the chemists who formulate the paint control the bonds between the droplets of oil, their size, and their number. It takes a lot of work to get the balance just right, which is why it's worth the premium you pay to get a good can of paint.

Even if you are not a painter and decorator, you will have come across non-Newtonian liquids in the kitchen. Like emulsion paint, tomato ketchup thins when it's under stress. It won't budge until you hit the bottle, putting the ketchup under enough sheer pressure for it to suddenly thin and shoot out of the bottle. That's why it's so hard to control the rate at which ketchup comes out of the bottle — if the force isn't high enough, it flows extremely slowly, but once you give it a big whack, the viscosity suddenly drops and it splatters all over your plate.

One of the most dangerous kinds of non-Newtonian behavior occurs when you mix together sand and water, creating a substance that's often called quicksand. Quicksand has semi-solid properties until put under pressure, and then it thins out, turning it into a fluid liquid — so-called liquefaction. That's why when you step into quicksand, the more you struggle and wriggle to get out, the more the liquid thins and the deeper you sink. But no matter what you see in the movies, most likely you won't die by sinking into quicksand; because it's a liquid with a higher density than your body, once you're submerged to your waist, you'll float back up. Still, getting out is very hard, since if you don't move, the liquid thickens and solidifies around you, and if you do struggle, it thins, making it hard to get a solid foothold. In other words, you're stuck until you're rescued — and that's when it can get deadly.

But more dangerous than quicksand is the liquefaction that occurs during earthquakes. Here, in another deadly example of non-Newtonian flow, the stress from the earthquake's vibrations liquefies the soil, usually causing massive damage. Just look at the 2011 earthquake in New Zealand: it struck the city of Christchurch, causing significant liquefaction that destroyed buildings and spewed thousands of tons of sand and silt onto the city.

As it turned out, non-Newtonian thinning was exactly the property that László Bíró needed to make the thick newspaper inks work in a fountain pen. He hypothesized that it would allow the ink to flow easily while you were writing, but then, once the ink was on the page, it would become thick and viscous again and dry into a solid so rapidly that it wouldn't smudge. Bíró started trying to make the perfect pen with his brother, who was a chemist; after many struggles, including having to immigrate to Argentina at the outbreak of World War II, they finally had something that worked. Their pens have a reservoir of ink that feeds a tiny rotating ball; when you write with the pen, the ball rotates, putting the ink under enough pressure to change its viscosity, so it comes flowing out onto the ball. At that point, the ink goes back to being sticky and gooey, until it hits the paper and flows out again. When you lift the pen, relieving the ink of its stress, it becomes thick again, and the solvents in the ink, which are being exposed to air for the first time, quickly evaporate, leaving the ink's dyes on the paper and creating a permanent mark. Genius!

As you might expect, over the years the ingredients used in

making such a high-performing ink have become trade secrets, but if you want to get a sense of just how good they are, write with a ballpoint pen on a piece of paper and then try to smudge the words with your finger. It's really hard to do. But that's not the only advantage that the non-Newtonian ink in ballpoint pens has over the more fluid inks in fountain pens. Because it doesn't flow under capillary action, the ink doesn't bleed as it seeps into the paper, as ink from other pens does. It's been chemically formulated to have a low surface tension when it comes into contact with cellulose fibers, as well as with the ceramic powders and plasticizers (known as sizing) that are added to the top surface of paper to make it glossy. Fountain-pen inks and other fluid inks have a high surface tension with sizing, so the ink sits on top of it and breaks up into small droplets. If you've ever tried to make notes on the front of a glossy magazine with a fountain pen, or tried to sign the back of a credit card with one, you'll have noticed this—the ink doesn't stay put. But ink from ballpoint pens seems to dry anywhere and stay exactly where you want it—even if you write upside-down— because it's not flowing, thanks to the force of gravity, but instead being rolled onto the page.

If you do try to write upside-down, you'll discover still another advantage of the ballpoint pen. Just like the fountain pen, it won't work if a vacuum forms in the reservoir of ink. But it has a simple way of preventing that—the top of the reservoir is open to the air, and the ink is quite viscous and won't flow without experiencing a lot of stress, so it doesn't fall out. Neat, eh? All that means is that, happily for the forgetful among us, you can leave a ballpoint pen at the bottom of your bag for months on end and it won't leak and cover your stuff with ink. Even if you forget to put the top back on and the ballpoint's left sitting unprotected in your pocket, the ink won't come out.

So good is this concept, and so reliable is the ballpoint pen at writing, even when the top has been off for months, that early manufacturers realized they didn't really need to put a top on the pen at all. Why not just retract the reservoir and ball back into the main body of the pen when you're not using it? That's easy enough to do, and so the retractable ballpoint pen was born. Click it, and you can write; click again, and the ballpoint is retracted. Oh, how the caliph of the Maghreb would have rejoiced at the sheer non-messiness and the audible delight of the retractable ballpoint pen!

The Bíró brothers produced the first commercial ballpoint pen while living in Argentina. They sold tons of them, to any number of clients, including the Royal Air Force, whose navigators snapped them up to replace the fountain pens they had been using, which always leaked at high altitude. Remembering this, I looked with renewed respect at the ballpoint I was holding in my hand — pilots and their crews were among the first to appreciate its brilliance, and I was happy to be filling in a customs form at altitude, using a descendant of those early ballpoint pens. According to the biggest manufacturer of ballpoint pens on the market today, the French company Bic, more than 100 billion pens have been made since they were first invented.

László Bíró died in 1985, but his legacy lives on. In Argentina they celebrate Inventors' Day every year on the anniversary of his birthday, September 29, and to this day, in the UK we call a ball-point pen a biro.

Of course, despite its success, many people hate the ballpoint pen. They decry its invention, saying it defiled the art of handwriting. It's true that the price for creating a portable, anti-smudge, anti-leaking, long-lasting, inexpensive, socially inclusive pen has been that the thickness of the line it creates cannot vary. Line thickness is determined by the size of the ball bearing at the tip, and because a ballpoint's ink doesn't flow once it's deposited on paper, the thickness of its line won't change by slowing down or speeding up your writing, as it would with the fountain pen, or other pens using Newtonian ink. Writing from ballpoint pens is more utilitarian, less expressive of an individual's writing style. But personally, I think it ranks up there with the bicycle for its impact on society. It's a piece of liquid engineering that's solved an age-old problem, has produced something utterly reliable, and is available at a price that is so affordable, most people regard ballpoint pens as communal property.

By the time I'd finished the customs form, I was so in awe of my ballpoint that I couldn't just put it back in my bag and consign it to another few months of being ignored. As I was trying to decide what to do with it, I realized that Susan was looking at me — a different Susan from the one I'd spent the flight with; a smiling Susan. She had her customs form in front of her, and she motioned to me, putting her thumb and forefinger together — miming the act of writing — and asking if she could borrow my ballpoint.

11 / CLOUDY

BONG, WENT THE INTERCOM, followed by a cabin announcement: "Ladies and gentlemen, as we begin our descent into the San Francisco area, please make sure your seat backs and tray tables are in their full upright position, that your seat belt is securely fastened, and all carry-on luggage is stowed underneath the seat in front of you or in the overhead lockers. Thank you."

The plane was descending now, and my ears were beginning to pop. I had that feeling of anticipation — that my life was going to start again after the suspended animation of airplane flight. This journey had pressed the pause button on my life, in exchange for a taste of omnipotence. Up here, the clouds couldn't rain on me; they couldn't tyrannically blot out the light and sway my mood as they do at home in London. Up here, the light streamed in through the window, warming my face with a glow from a sun that never set. Never, that is, until the plane suddenly descended into the cloud layer; then, not only did the sun disappear, but it was abruptly replaced by a white haze that knocked all sense of omnipotence and security out of me: white out!

The cloud we had descended into, like all clouds, was comprised of liquid droplets of almost pure water. The almost bit is interesting; it's the reason why rainwater isn't pure, why windows get stained by rain, and why fog forms in some places and not others. The water in clouds is neither pure nor innocent — it can kill. Night and day, somewhere on the planet, lightning storms are raging, at a pretty constant rate of fifty lightning flashes per second, globally. It's estimated that there are more than a thousand human deaths by lightning per year, with the injured numbering in the tens of thousands. The US National Weather Service keeps a running total of the deaths, and the particulars of the fatalities. The table shows some of the entries for 2016. You'll see that taking shelter under a tree is not a good idea, and that danger can strike almost anywhere. But can lightning get to you on a plane? That is a question worth answering.

Clouds start off as wet laundry on a line, as a puddle on the pavement, as a glow of perspiration on your upper lip, as part of a vast ocean of water. Every second, some of the H_2O molecules leave wet laundry, puddles, upper lips, oceans, and other bodies of

Day	State	City	Age	Sex	Location	Activity
Fri	LA	Larouse	28	F	In tent	Attending Music Festival
Fri	FL	Hobe Sound	41	M	Grassy Field	Family Picnic
Fri	FL	Boynton Beach	23	M	Near Tree	Working in yard
Wed	MS	Mantachie	37	M	Outside Barn	Riding Horse
Wed	LA	Slidell	36	M	Construction Site	Working
Mon	FL	Manatee County	47	M	Farm	Loading Truck
Fri	FL	Daytona Beach	33	M	Beach	Standing in water
Sat	MO	Festus	72	M	Yard	Standing with dog
Mon	MS	Lumberton	24	M	Yard	Standing
Sun	LA	Pineville	45	M	Parking Lot	Walking to car
Thu	TN	Dover	65	F	Under tree	Camping
Thu	LA	Baton Rouge	70	M	Sheltering under tree	Roofing
Thu	AL	Redstone Arsenal	19	M	Outside building	Outdoor maintenance
Thu	VA	Bedford County	23	M	Along roadway	Walking
Sat	NC	Yancey County	54	M	Putting on rain gear	Riding Motorcycle
Tue	CO	Arvada	23	M	Sheltering under tree	Golfing
Tue	AL	Lawrence County	20	M	In yard under tree	Watching Storm
Wed	AZ	Coconino County	17	M	Near mountain top	Hiking
Fri	UT	Flaming Gorge	14	F	On reservoir	Riding jetski

A table of deaths caused by lightning in the United States, collected by the National Weather Service.

water, and make their way into the air. The boiling point of water is 212°F, denoted as the temperature at which pure liquid turns into a gas at sea level. So how does liquid water become a gas without reaching this temperature? What's the point of defining the boiling point if water can cheat and dry laundry and upper lips, evaporate puddles, and denude oceans autonomously, at much lower temperatures?

As it turns out, the definitions of solids, liquids, and gases are not as clear-cut as they might seem, and the game that scientists play, of categorizing the world and making neat distinctions between different things, is constantly being sabotaged by the complexity of the universe. To understand how water cheats the system to create clouds, we have to think about an important concept called entropy.

The water clinging to your laundry on a clothesline is below 212°F in temperature, but it is in contact with air. The molecules in the air bombard your washing, crashing into it as they move chaotically; occasionally, in all the mayhem, an H_2O molecule pings off to become part of the air. It takes some energy to do this, as the bonds that attach the H_2O molecules to your wet clothes have to be broken. Taking the energy away from your clothes cools them, but it also means that if the H_2O molecule floating around in the air were to collide again with your washing, it would gain energy by sticking to it, thus making it wetter again. So on average you might think that more water would stick back onto your clothing than would be carried away by the air currents of the wind. But here's where entropy comes into play. Because the amount of air billowing around your laundry is so great, and the number of water molecules is so low, the chances of a water molecule finding its way back onto your favorite cotton T-shirt is small. Instead, it is more likely to be whizzed up into the atmosphere. This propensity of the world of molecules to get jumbled up and spread out is measured by the entropy of the system. Increasing entropy is a natural law of the universe, and it opposes the forces of condensation that bond the water back onto your washing. The colder the temperature and the less exposed your laundry is to the wind, the more you tip the balance in favor of condensation, and your washing stays wet. In contrast, by hanging your washing on a line on a warm day, you tip the balance in favor of entropy, and your clothes get dry.

Entropy also takes care of puddles in the street, dries your bathroom after you've been in the shower, and removes the sweat from your body on a hot day. All in all, entropy seems very convenient, and generally quite helpful, given how much we like having dry clothes and bathrooms, and cool bodies. But that same benevolent force also drives the killer clouds that strike us down in the thousands every year by throwing their lightning bolts around, reminding us who's really the boss in our atmosphere.

The process of thundercloud formation starts with vaporized H_2O, which moves around as a gas. Hot air rises because it's less dense than cold air, so on a sunny day the water molecules make their way from your washing up into the atmosphere. The air, though full of water, is transparent, so at first there won't be any sign of a cloud. But as the vapor goes higher, the air expands and cools, and the thermodynamic balance is tipped toward H_2O molecules preferring to condense and be part of a liquid again. But a single molecule can't just turn back into liquid in midair; to form a tiny droplet of water requires some coordination—several H_2O molecules all have to come together to become a single droplet. In the chaotic, turbulent atmosphere, this doesn't happen easily, but the process is expedited by the tiny bits of particulate matter that are already in the air—often small bits of dust that have blown off trees and plants, or smoke from factory chimneys. The H_2O molecules can attach themselves to these, and as more and more of them join together, the particle becomes the center of a tiny droplet of water. This is why when you collect rainwater, it generally contains sediment, and why when the rain dries onto your car windshield or the windows of your house, it leaves a fine powder.

This piece of physics was at the heart of one of the most extraordinary experiments of the twentieth century—when scientists took it upon themselves to control the weather. The method was called cloud seeding, and it was invented in 1946 by Vincent Schaefer, an American scientist. Schaefer and his team determined that if you dispersed silver iodide crystals into the atmosphere, they would act as dust or smoke might, and become the nucleating droplets—the seeds—of clouds, which would, in turn, produce snow and rain. The technique is an art as much as it is a science, but as widely used as it's been for decades, many dispute its effectiveness.

Still, though, the USSR seeded clouds over Moscow every year. Their aim was to clear moisture from the air by making it rain, ensuring that their May Day celebrations were accompanied by blue skies. The US military employed the technique for a different aim during the Vietnam War—they used it to extend the monsoon season on the Ho Chi Minh Trail; this was called Operation Popeye, and its mission was to "make mud not war." Today countries around the world, like China, India, Australia, and the United Arab Emirates, all experiment with cloud seeding as a means to tackle drought conditions. Of course, by seeding the air you control only one aspect of the weather: cloud formation. So if the moisture content of the air is low, no amount of cloud seeding will make it rain. But if the air is full of water, then using this technique to increase snowfall over ski resorts, or reduce the risk of hail damage on crops during storms, can be productive. In the aftermath of the Chernobyl nuclear disaster in 1986, cloud seeding was used to make enough rain to remove radioactive particles from the atmosphere.

Airplanes don't need to use silver iodide to seed clouds. If you look up into the sky on a sunny day, you'll often see contrails emanating from the back of a jet aircraft. This isn't smoke spewing out of a badly maintained engine; it's a cloud seeded by the engine emissions. Small particles from the combustion process are emitted from the plane, along with an enormous amount of very hot gas. The gas pushes the aircraft forward, and while you might expect it to be too hot for water to form, at high altitudes the temperature is so low that the exhaust is quickly cooled. The emission particles become sites of nucleation for liquid droplet formation, which then freeze, first becoming water, and then tiny ice crystals. Contrails are just high, wispy cirrus clouds.

Depending on conditions in the air, contrails might last for just a few minutes, or maybe for a few hours, and the sheer number of them (there are a hundred thousand flights per day, globally, all producing contrails) has led many to suspect that contrails must have an effect on Earth's climate. Common sense tells you that clouds cool our planet; if you've sat on the beach on a cloudy day, you'll have experienced this. But clouds don't just reflect sunlight back into space. They also trap heat from the ground, in the form of infrared waves, and bounce it back to Earth. It's an effect that's particularly noticeable in winter, when clear skies create colder conditions than cloudy skies, because at night the heat that's lost from the ground is bounced back by clouds. And different cloud types (distinguished by color, density, and size) at different heights have different effects. All of which is to say, determining whether contrails have a net warming effect or a net cooling effect on Earth's average temperature is an outstanding scientific question.

Investigating this question requires being able to study Earth's climate in the absence of contrails and compare average temperatures with and without. But there are always aircraft flying somewhere in the stratosphere. When planes land for the night in America, they are just taking off in the Far East and Australia, and when they stop flying there, European planes take off, and so it goes — it's a 24/7 global operation. There are more than a million people in the air at any given moment. The only time this was not true in recent memory was after the terrorist attack on the Twin Towers in New York. All planes were grounded in the United States for three days after September 11, 2001. The measurements from four thousand meteorological stations across the United States showed that on 9/11 the difference between daytime and nighttime temperatures was, on average, 1.8°F higher than usual. This, of course, is just one study and at one time of year, autumn. It's quite possible that in winter, spring, and summer, when cloud coverage and the localized climate would be different, the net effect of contrails would be to decrease temperatures, not increase them. There's a lot of ongoing work in this area, but it will never be an easy matter to resolve; our climate is complex. Certainly, it's hard to imagine a time when we'll be able to collect more data from a complete no-fly scenario, given that flying is such an important part of global culture. Nevertheless, scientists have widely discussed the possibility of controlling global temperatures through seeding clouds, and whether that would have the potential to avert some of the effects of climate change. Many suspect that they could manage solar radiation by increasing the reflectivity of the atmosphere by making clouds whiter.

The deliberate manufacture of contrails seems like an obvious way of testing that theory, and although such experiments are highly controversial, there are some who think they're already being carried out in secret. Contrail conspiracy theorists argue that some contrails stay in the sky for too long, and that the only way that could be happening is if they're being created by aerosols and other chemicals. Some of the conspiracy theorists go further still and argue that the contrails are evidence that governments have been spraying liquids across their territories, with the aim of psychologically manipulating the population through chemical means.

These conspiracies play off legitimate fears that we could be manipulated and poisoned through the water we drink. The danger is real; the water supply has historically been responsible for the mass poisoning of whole communities. It still goes on in modern times; for instance, as recently as 2014, the entire city of Flint, Michigan, was poisoned by lead in the water due to government incompetence. The outbreak of cholera in Yemen, which began in 2016 and is now nearing one million cases, was caused by the breakdown of a clean water supply. Not surprisingly, the fear of mass infection and poisoning has been a common motif in fiction, perhaps most famously in the film *Dr. Strangelove,* in which General Jack D. Ripper identifies the fluoridation of water as a Communist plot to undermine the American way of life: "I can no longer sit by and allow the international Communist conspiracy to sap and impurify all of our precious bodily fluids," says General Ripper, before initiating a nuclear attack on the USSR.

This movie, perhaps the greatest film to examine the circumstances under which a nation might initiate nuclear war, is right to identify the adulteration of water as a potential motive for global conflict. We all need clean water to drink — we cannot live without it — and if our water is adulterated or contaminated, it will bring about death and disease on an epic scale. The pandemics of cholera in the nineteenth century killed tens of millions of people before anyone understood that the disease was brought on by waterborne bacteria.

And as with all liquids, water is very hard to control. It goes everywhere, eventually, moving from lakes to rivers to oceans and up into the skies. Thus the fear of water contamination is as great today as ever, and yet the water coming down from the clouds, the origin of most of the water we drink, is equally difficult to protect. Clouds know no territorial boundaries; one nation's experiments, disasters, or actions can and do affect the rest of the world in the most intimate way.

Dr. Strangelove was made as a satire, but the suspicion and fear surrounding the potential contamination of what we put into our bodies are real and will probably never go away. "Foreign substances introduced into our precious bodily fluids without the knowledge of the individual, certainly without any choice, that's the way your hardcore Commie works," says General Jack D. Ripper. But replace "Commie" with "federal government," or "capitalist corporation," or "scientist," or even "environmentalist," and you have the essence of most arguments against any number of policies concerning vaccination, water chlorination, or even power generation. There are countless examples of this — just look at acid rain.

Coal often contains impurities in the form of sulfates and nitrates, which become sulfur dioxide and nitrogen oxide gases when the coal is burned. These gases rise and become part of the atmosphere, and then dissolve in the liquid droplets that make up clouds. The presence of these gases makes those droplets acidic, so when they return to earth in the form of rain, they acidify the rivers and lakes and soil, killing fish and plants, and destroying forests. Acid rain also corrodes buildings, bridges, and other infrastructure, and it often does so very far away from where the original emissions — the gases from the coal — came from. It falls in a different country from where it was initially emitted, becoming a political issue as well as an environmental one. The cause of acid rain was identified during the industrial revolution in the nineteenth century, but it wasn't until the 1980s that the West, the main producer of acid rain, made a concerted effort to combat it.

The nuclear disaster at Chernobyl in Ukraine in 1984 caused another pan-national problem carried by clouds. When it became clear that the radioactive elements from the explosion at the power station had become airborne, everyone knew that the prevailing winds would determine which countries would be affected. The UK was one of those countries, with sheep farmers in England and Wales suffering from radioactive rain falling on their land, becoming a part of the soil and grasses. If swift preventive measures had not been taken to stop sheep from eating this grass, they would have become radioactive too. It was only in 2012, twenty-six years after the Chernobyl explosion, that restrictions were finally lifted by the UK Food Standards Agency on sheep being raised in the affected regions.

The world is a connected place. It's connected through the clouds and the rain showers they produce, and in another sense, it's connected through airplane travel. As I gazed out the window into the whiteness, I found it hard to reconcile a cloud as being fundamentally liquid. The individual droplets that make up a cloud are of course too small to see, but they are also transparent. So why are clouds white?

Well, while light from the sun passes straight through many of the droplets in a cloud, sooner or later it will hit a droplet and be reflected, just as the sun is reflected off the surface of a lake. This bounces the light off in another direction, so it hits another droplet and is reflected again. This continues, and the ray of light is bounced around like a pinball until it leaves the cloud. When it finally reaches your eyes, you see a pinprick of light originating from the last droplet of water that the light bounced off. The same happens to all the other rays of light that hit the cloud, so that what your eye sees is billions of pinpricks of light originating from all over the cloud. Some will have taken longer routes and lost their brightness, and so that part of the cloud will appear darker. Your brain tries to make sense of all these pinpricks of light. It is used to interpreting shades of light and dark in reference to a three-dimensional object, which has material characteristics that correlate with what you are seeing. This is why clouds appear to be objects, sometimes fluffy as if made of wool, and sometimes denser, as if they might be a floating mountain. Of course, another bit of your brain denies all this, and points out to your subconscious that these are not objects at all, but tricks of the light. Still, even knowing this, it's hard to see clouds as just an agglomeration of water droplets.

Much of the beauty of the sky is due to clouds and their water content. It affects the light we perceive in myriad ways and is one of the main reasons why different places in the world are so sublimely different in terms of light. But as the tiny droplets that make up a cloud become more dense, it becomes harder and harder for light to bounce its way through from top to bottom, and the cloud appears dark gray. We all know what this means, especially in Britain—it's going to rain. The tiny spherical water drops that are floating in the cloud start to get bigger, and gravity begins to exert a greater force on them. When the droplets are just the size of tiny dust particles, buoyancy and air convection currents exert a far greater force on them than gravity does, so they just float around like dust. But as they get bigger, gravity starts to dominate, pulling them down toward Earth and turning them into rain. If we're lucky, that is; otherwise they may form a storm cloud, the very storm clouds that kill hundreds of people every year.

Storm clouds are made under a very particular set of circumstances. As droplets experience cold air, water vapor changes from a gas back into a liquid. It's the opposite of what happens when your wet clothes dry on a clothesline. In doing so, it gives off energy in the form of heat—we call this latent heat. Latent heat is emitted from H_2O molecules while they're still inside the cloud, meaning the air in the cloud gets warmer. As we know, warm air rises, so the cloud bulges out at the top. That's how puffy cumulus clouds are made. But if all that happens while a lot of warm, humid air is rising up from the ground—as might happen on a summer's day—then the convection currents pushing the cloud droplets upward might be strong enough to reverse the rain and send it upward too; the droplets will go miles into the sky, until the air carrying them finally cools enough to stop rising. That high up in the atmosphere, the rain droplets freeze, become ice particles, and then fall again, but depending on the climatic conditions, they might be pushed upward again by still more warm air. Meanwhile the cloud is getting bigger and taller, and darker and darker, a cumulus cloud being transformed into a cumulonimbus cloud —a storm cloud. The convection currents pushing the droplets up increase to speeds of sixty miles per hour, and the cloud becomes a complex swirl of activity, with ice particles falling through an updraft of air that's carrying still more droplets, all of which are colliding violently over several miles.

The scientific community still isn't sure how the conditions inside a cumulonimbus cloud lead to the buildup of electric charge. But we do know that, as it does on the ground, the electricity arises due to the movement of charged particles, which originate from atoms. All atoms share a common structure: a central nucleus containing positively charged particles called protons, surrounded by negatively charged particles called electrons. Occasionally some of the electrons break free and start moving around; this is the basis of electricity. When you rub a balloon on a wool sweater, you create charged particles on the balloon. Then, if you hold that balloon up to your head, your hair will move in response to the charges on the balloon, attracting the opposite charges on your hair. Negative charge ultimately wants to be reunited with positive charge, and it stretches your hair toward the balloon in order to accomplish this, causing your hair to stand on end. If the amount of charge were higher, there would be enough energy for the charged particles to jump through the air, creating a spark.

In a cloud, instead of gently rubbing balloons you have water droplets and ice particles, all turbulently crashing into one another with tons of energy, giving some of the ice particles a positive charge as they're carried to the top of the cloud, and some of the raindrops a negative charge as they fall to the bottom. This separation of positive and negative charges over many miles of cloud is driven by the energy of the winds inside the cloud. But the attractive force between the positive and negative is still there — they want to get back together, which is to say there is a voltage building up inside the cloud. It can get so large, reaching hundreds of millions of volts, that it strips electrons away from the molecules in the air itself. When this occurs, it happens very quickly, triggering the release of an electric charge that flows between the cloud and Earth, or between the top and the bottom of the cloud, depending on the conditions. The discharge is so big that it glows white-hot — it's lightning. And thunder is the sonic boom of the surrounding air rapidly expanding as it's heated to tens of thousands of degrees in temperature.

The energy of lightning is so huge that it can and does vaporize people, hence the high death toll. Electricity always flows down the path of least resistance—it's like a liquid in that respect. But while liquids flow down gravitational fields, electricity flows down electric fields, and since air doesn't conduct electricity very well, it has high resistance to the flow of electricity. Humans, on the other hand, are composed mostly of water, which does conduct electricity well. So if you're a lightning bolt emanating from a thundercloud, trying to find the path of least resistance to Earth, a person is often your best vehicle. While lightning might prefer to go through a tree because it's taller and longer, and thus more of the conductive path can go through its watery branches, if a person is sheltering under that tree, then the lightning might, and often does, jump over to the person on the last part of its journey to Earth.

Throughout a lot of the world, the tallest structures are often buildings, and in the West, for a long time, the tallest building in any town or city was a church. Many early church spires were made of wood, and they would burst into flames when lightning hit them. Fortunately, in 1749, Benjamin Franklin realized that if you just placed a metal electrical conductor on top of buildings and connected that to the ground with a piece of conducting wire, you'd give the lightning an easier path down and thus avoid a lot of the destruction caused by lightning strikes. These conducting wires are still used today and continue to save hundreds of thousands of tall buildings from being damaged by lightning. The same principle explains why being inside a car protects you from lightning: if the lightning strikes the car, it will be conducted around the outside of the metal bodywork, a path less resistant than going through the passengers.

Which brings us to aircraft and the dangers of lightning. When an aircraft is flying through a storm cloud, the turbulent air causes the aircraft to shake and roll, to drop or rise suddenly as the pressure changes. If, in the midst of this, there's lightning in the clouds, the plane will most likely become a part of that lightning's conductive path. As we know, many older aircraft are built from an aluminum alloy fuselage, and, as it would in a car, the metal protects the passengers from the lightning's charge. But the carbon fiber composites that modern passenger aircraft are made of don't conduct electricity very well (the epoxy glue holding the carbon fibers together is an electrical insulator), so, to compensate for this, aircraft-grade carbon fiber has conductive metal fibers built into its composite structure, ensuring that when lightning strikes, it travels around the skin of the aircraft and doesn't harm the passengers. So while aircraft get struck fairly frequently — once a year on average — there haven't been any recorded accidents on planes as a result of lightning strikes in over fifty years. In other words, it's more dangerous to be on the ground, under a tree, during a lightning storm than in an airplane. They don't mention this in the preflight safety briefing, even though it makes flying much safer. But — as already discussed — the preflight briefing is not really about safety.

By now my plane was getting pretty close to the ground, relatively speaking. As we continued to reduce altitude on our approach to San Francisco International Airport, the low clouds kept us from seeing much out the window. The San Francisco Bay Area is prone to fog. Fog, like clouds, is a liquid dispersion of water droplets in air: it's essentially a cloud at ground level. Fog seems harmless enough if you're looking out at it from a snug house, warmed by a log fire, sipping a glass of brandy — it gives the city an air of romanticism, a sense that mysterious new things are possible. But if you're walking on a moor or driving on a highway or skiing down a mountain or descending at a rate of thirty feet per second in an airplane, fog means just one thing — potential death. The history of sea fogs and ships dashed upon rocks as a result of the crew's not being able to see these hazards is still a very real and frightening part of seafaring life. Fog can close airports and can cause airplanes, unless they are fitted with modern fly-by-wire systems, to abort landings. Fog is scary, fog is dangerous — and perhaps that's why celebrations of the dead, like Halloween, are often held at times of the year when fog and mist are prevalent.

Fog forms at ground level for the same reason that clouds form in the sky. Damp, humid air cools, and as a result the H_2O in the air liquefies into fine droplets of water. Just like at high altitudes, the formation of droplets requires a site of nucleation, and traditionally, in cities, that was provided by smoke from fires that were used for cooking or for keeping houses warm. But in modern times the nucleation sites usually stem from the exhaust from factory chimneys and cars. When there's a chronic excess of that kind of pollution, a thick fog called smog will form, often hovering for days on end, capturing the pollution and keeping it above the city. In London, the recorded history of smog goes back to 1306, when King Edward I banned coal fires for a period of time to try to combat the problem. The smog got so bad, at times you couldn't see your hand in front of your face. But despite Edward's efforts, smog continued to plague London for centuries, until the Great Smog of 1952, which was so lethal that it killed four thousand people in just four days, provoking the government to pass the country's first clean-air laws.

San Francisco often experiences dense fog. This is due to a combination of conditions that bring the warm, damp air of the Pacific Ocean over the city, where it then cools and condenses into fog as a result of car exhaust emissions. We were descending into just such a fog now, and despite knowing that those at the controls of the aircraft and on the ground at the airport are accustomed to these conditions and know how to make a safe landing in them, I felt myself becoming increasingly anxious as we continued to descend toward the ground while outside there was nothing to see but white spookiness.

Bong went the intercom. "Flight attendants, please prepare for landing." The safety-critical moment had arrived — we were going in. The cabin fell silent, except for the drone of engines and the blast of the air conditioning. Everyone seemed to be tuning in to the same anxiety. Occasionally the fog would clear enough for me to catch sight of some feature on the ground, a tree or a car, but then the whiteness would reassert itself, and the aircraft would wobble or drop as the engines warbled away into my fearful ear.

As we got lower and lower, I felt more and more tense. I know, rationally, that flying is the safest form of long-distance travel, but I'm always worried about being the exception. The deadly fog was outside. We were all strapped in, including the cabin staff, who were looking out at us impassively. They did this several times a week. *How did they cope,* I wondered, *with this last part of the flight, this part when it's clear that our lives are in the hands of the pilots' ability to cope with something unseen and unexpected?* Only supercool Susan seemed unaffected; she'd put her book down and was looking out of the window serenely, clearly completely confident that our imminent collision with the ground would be a success.

12 / SOLID

THERE WAS A THUD, and the whole fuselage shuddered, making the sound of a thousand cupboards being shut. We all lurched forward and strained our seat belts as the captain cut the power on the jet engines and we decelerated from a landing speed of 130 miles per hour to 70, then to 40, then to 15, as we taxied off the runway. There was audible relief in the cabin; a few people clapped: we were back on solid ground.

Although *solid* is not really the right word, since, so far as planets go, Earth is not particularly solid. It started as a ball of hot liquid, and over the course of a hundred million years cooled enough for a thin crust of rock to form on the outside. That happened about 4.5 billion years ago, and our planet has been cooling ever since, but it's still fluid on the inside. It's the dynamic flow of liquid inside Earth that keeps our planet alive by creating our protective geomagnetic field. But this same fluidity is also a destructive force, causing earthquakes, volcanic eruptions, and the subduction of tectonic plates.

Right at the center of Earth there is indeed something solid — a core of metal made of iron and nickel, at a temperature of more than 9,000°F. It's solid even at this temperature, which is thousands of degrees above its normal melting point, because the intense gravitational pressure at the center of Earth forces the liquid to form giant metal crystals. The core is surrounded by a layer of molten metal, again mostly iron and nickel, more than a thousand miles thick. The currents flowing inside this interior metal ocean are what produce Earth's magnetic field, which is so powerful that it extends outward, not just to the surface, where its force makes compasses work, allowing us to navigate, but also far into space. Out there, Earth's magnetic field acts like a shield, performing a vital role in protecting us from the solar wind and cosmic rays raining down on us. Without our magnetic shield, they would strip us of our atmosphere and water, and most likely kill off all life on the planet. Planetary scientists believe that Mars lost its magnetic shield some time ago, which is why it has no atmosphere and has become a cold, dead planet.

Surrounding our ocean of liquid metal is a layer of rock between 1,000°F and 1,500°F — the mantle. At these red-hot temperatures, the rock behaves like a solid over time periods of seconds, hours, and days, but like a liquid, over periods of months to years. This is to say the rock flows even though it's not molten — we call this type of flow "creep." The major flows of this rocky mantle are convection flows: the hot rock near the liquid metal ocean rises up, and the colder rock closer to the crust sinks down. This is the same kind of flow you can see in a pan of water while it's heating up; the hot water at the bottom of the pan expands and becomes less dense than the colder water at the top of the pan, which sinks down to replace it.

On top of the mantle is the crust, which is like the skin of Earth. It's a relatively thin layer of cool rock, between twenty and sixty miles thick, and it's covered by all of the planet's mountains, forests, rivers, oceans, continents, and islands. And as the intercom went off once again, our flight attendant confirmed that we'd just landed on it.

"Ladies and gentlemen, welcome to San Francisco International Airport, where the local time is 3:42 p.m. and the temperature is thirty-seven degrees. For your safety and comfort, please remain seated, with your seat belt fastened, until the captain turns off the FASTEN SEAT BELT sign."

At moments like this, the relief of being back on the ground might make you feel that the crust on which we live is a stable solid, which we can steadfastly rely on. Unfortunately this is not the case; the crust is essentially floating on the fluid mantle below, and to make the whole thing still more precarious, it's made up of separate pieces, called tectonic plates. The mantle's convection forces move the tectonic plates around, causing them to buckle as they bump into one another. There are seven major tectonic plates, which generally line up with the continents — so, for instance, the North American tectonic plate contains North America, Greenland, and all of the seafloor between there and the Eurasian tectonic plate, which contains most of Europe. All the tectonic plates move, but not in the same direction, and the places where they meet, called fault lines, are collision zones. As the plates push together, they rise up to form mountains. Where the plates pull apart, new crust is formed, as lava shoots up from the mantle below. The fault lines are also where the most violent earthquakes occur.

I'm sure my fellow passengers understood the danger — how could they not, if they lived in a place like San Francisco? The city is located at the fault line where the North American tectonic plate meets the Pacific tectonic plate, and so has a long history of major earthquakes, and there will surely be more to come. In 1906, an earthquake destroyed 80 percent of the city and killed more than three thousand people. There was another one in 1911, then another in 1979, and others in 1980, 1984, 1989, 2001, and 2007. And these are just the big earthquakes. There have been many more smaller disturbances in the crust in that time. Living in a place like San Francisco makes it clear how important it is to understand the fluid dynamics of our planet. Not only does it explain why massive earthquakes occur and reoccur in certain places, but it also helps us understand the factors that affect a vitally important related quantity: the sea level.

Because Earth's crust sits on top of fluid rock, if it's weighed down by, say, miles of ice, then it will sink into the mantle. This is what's happened to Antarctica and Greenland, which are both covered in one to two miles of thick ice. To get a better feel for the scale of these ice sheets, consider that the Antarctic ice sheet holds 60 percent of all the fresh water on the surface of the planet — approximately seven million trillion gallons of water, which weighs approximately twenty-eight thousand trillion tons. If global warming were to cause all that ice to melt, then the sea level of the oceans would rise by more than 160 feet, submerging every single one of the world's coastal cities and making hundreds of millions of people homeless. This seems obvious. What is less obvious is that the release of the weight of ice from Antarctica will de-stress the rocks underneath it, and those landmasses will decompress and bob up (this is called postglacial rebound). Greenland is in a similar situation: the crust below it is being weighed down by the million trillion gallons of water held in the ice sheet, and if all that melts, then the North American tectonic plate will rise. If the resulting increase in the height of the continent is greater than the rise in the sea level, then major flooding may be avoided. Working out what's more likely to occur is vitally important for our future, and especially for future generations, because if global warming intensifies, and it is well along the path to doing so, one of these scenarios will certainly play out.

At the moment, what we know is this. The mean global sea level has risen almost eight inches since the beginning of the twentieth century. Some of this has been due to the water thermally expanding as the oceans have got hotter, since hotter liquids take up more volume. Some of the rise has been due to the melting of the ice sheets over Greenland and Antarctica, and still more because glaciers in other parts of the world are melting too. The rising sea levels are global; they affect everyone with a coastline, from a tiny Pacific island, which will be entirely submerged, to a huge country like Bangladesh, where a three-foot rise in sea level would result in nearly 20 percent of the country being submerged, and thirty million people being displaced. On the other hand, the postglacial rebound affects only the coasts connected to the parts of Earth's crust weighed down by the ice sheets in Greenland and Antarctica. In other words, there will be winners and losers when Earth's ice melts, and it all depends on which part of it melts first: Greenland in the Northern Hemisphere, or Antarctica in the Southern Hemisphere.

If the ice melts in the Northern Hemisphere first, then Greenland will bounce up higher than the average sea level, as will the North American continent, so sea levels there will initially go down. The extra water will be distributed across all the oceans, while the increase in the height of the northern tectonic plates will be a local effect. If the opposite happens, and the Antarctic ice melts before the Greenland ice sheet does, then the southern tectonic plates will bounce up first, and the whole east coast of North America will be under water.

One of the big unknowns is how fast the ice will go, since it doesn't have to melt in order to disappear from continents. It can also creep: this is how glaciers move, flowing down mountains even though they are solid ice. How creep works is not so different from how viscous liquids ooze. When the force of gravity is applied to a molecule in a liquid, some of the weak bonds holding it together break, allowing it to move in the direction determined by the force. But it also needs to have space to move into, and if it can't find that, it exerts pressure on its neighboring molecules, prompting them to move. The structure of a liquid is mostly random, so spaces do often open up, allowing molecules to move and shuffle freely in response to forces, and the liquid to flow. The same thing happens in solids, but the molecules and atoms have relatively less energy to break the bonds holding them to their neighbors, so the process is dramatically slower. Solids also have a very orderly structure, so finding space for the atoms to move into is difficult. That's why they flow so slowly, and that's why we call it creep. You can speed up creep by putting solids under higher pressures, or by increasing their temperature; at higher temperatures, the atoms have more vibrational energy to break existing bonds and jump into whatever space may be available. This is what is happening to the ice sheets as global temperatures rise: whole mountains of ice are flowing, driven by gravity toward the sea.

In the form of glaciers, ice creeps relatively fast. In 2012, for instance, glaciers in Greenland were measured to be moving at a rate of ten miles per year toward the sea. They were moving so quickly because the ice sheets had reached temperatures between $-50°F$ and $-100°F$. As cold as that may sound, that ice is only $82°F–132°F$ below its melting point of $32°F$. Which is to say that the energy of the H_2O molecules inside the ice crystals is not very far from the temperature they need to be at in order to turn into liquid water. In contrast, the rocks that make up a mountain have melting points between $2,000°F$ and $4,000°F$, so the atoms in the rocks of a large mountain are thousands of degrees below their melting point, and behave much more like a solid than a glacier does. Thus, mountains creep more slowly than glaciers, but they creep nevertheless; it just takes them millions of years to flow appreciable distances. Lower down in Earth's crust, temperatures are closer to the melting point of the rocks, which is why tectonic plates creep faster than mountains, at rates from less than one to almost four inches per year.

That may not sound like much, but now imagine that there's another tectonic plate pushing against the rock, and that the forces are acting over a fault line hundreds of miles long. Something has to give. If not, then year after year, the tension will keep building up until the rock ruptures and slides, causing a nearly instantaneous, enormous release of energy—an earthquake. The amount of energy released in the 1906 earthquake in San Francisco was equivalent to that of about a thousand nuclear bombs. The earthquake that caused the tsunami that hit Japan in 2012 released energy equivalent to that of twenty-five thousand nuclear bombs. It's this gigantic output of energy that makes the damage from earthquakes so widespread; one big earthquake, with an epicenter hundreds of miles away from any city, can still be devastating.

But that buildup of energy doesn't always create an earthquake; sometimes the rock creeps and, like two pieces of paper being pushed together, it slowly flows upward to release the pressure. This requires a huge amount of force, but then a huge amount of force is exactly what's produced by tectonic plates. It's that inexorable wrinkling that makes mountains. The great mountain ranges — the Alps, the Rocky Mountains, the Himalayas, and the Andes — are all located where tectonic plates meet, and all were formed through creep, over millions of years.

But not all mountains are made this way. Perhaps the most impressive way of making a mountain, and certainly the fastest, is through volcanic eruption. If you've not seen hot rivers of red molten rock bursting from the bowels of Earth, you really should try to, at least once in your life; it's one of nature's most awesome and humbling sights, a bit like going back in a time machine to the birth of the planet, where everywhere you'd look, there'd be burnt rock and black cinder, accompanied by the smell of sulfur, smoke, and ash.

The only time I've witnessed a live volcano, I very nearly got killed. I was living for a brief time in Guatemala, studying Spanish; it was the summer of 1992 and I was lodging with a family in the old city of Antigua, located in a mountainous jungle region on the Central America Volcanic Arc, a chain of volcanoes on the Pacific coast all created by tectonic activity. Over the past three hundred thousand years, it's estimated that seventeen cubic miles of mountain have been built up by the eruptions of these volcanoes. One of the most active in the region is Pacaya, which is close to Antigua and had its last major eruption in 2010.

When I was in Antigua, guides in the market square arranged unofficial visits to the volcano. The Guatemalan family I was staying with had warned me against going because in 1992 the country was still full of bandits and outlaws who regularly robbed any tourists who were young and foolish enough to go out into the countryside unarmed. But because I was young and foolish, I took no notice of their advice, and so off I went late one afternoon, in a truck full of equally young and foolish backpackers, driven into the jungle by two young Guatemalans. We got to the base of Pacaya as the sun was setting and began our climb up through the trees—only there weren't any trees because Pacaya, an active volcano, intermittently erupts, sending up plumes of smoke and ash and lobbing tons of molten rock into the air. These emissions had burned and destroyed all of the forest that had once grown around the cone, so where we were, at the bottom, there was just a steep slope of ash, punctuated every ten yards or so by blackened stumps. As we started hiking, we walked along this black mound of loose cinder, with foul-smelling smoke drifting all around us. It felt like a scene from the apocalypse. As we continued upward, the path got steeper, and it became harder to make progress through all the cinder. But we were eager and adventurous, and finally we made it to the top, just as darkness fell.

Soon it was pitch-black, and our guides motioned for us to stay behind a big rock near the edge of the crater while they moved forward to see what kind of mood Pacaya was in. They came back quickly, excited to tell us that she was awake and bubbling with lava. So we crept forward too. The smell of sulfur billowed up from the crater, which seemed to be somewhere between a hundred and two hundred yards below—I couldn't really tell. And then we saw the lava. It was one of those moments I will never forget—like seeing the inside of our planet for the first time. We were all transfixed, as if we were watching some wild animal in its lair. It was then that we heard some popping sounds. Our guides suddenly looked worried and conferred privately. There were more pops and some faint thuds. It seemed that Pacaya really was awake, and was shooting molten lava into the air—the thuds were the sounds of lava landing. Each spurt, I found out later, could have been two to four pounds in weight. We weren't wearing safety helmets, or heat-resistant clothing, or even boots (I was wearing sneakers). The guides told us the best thing to do in this situation was just to run, and we didn't need any persuading. I fled, terrified that the next pop would land a splat of molten lava on my head, and so I slid, and fell, down the mountain of cinders as fast as I could, all the while hearing the *pop, pop, pop* behind us. In the truck back to Antigua, our guides laughed—that was a close one, apparently. I finally understood why they weren't worried about bandits; they were a danger for sure, just not the biggest one.

But in the grand tectonic scheme of things, Pacaya's eruptions are minor. The planet's biggest volcano is Mauna Loa, on the Big Island of Hawaii, which it created with its magma. Most volcanic activity occurs under the sea. The Hawaiian Islands were all built by volcanic activity, and that continues today, making them a pretty dangerous place to live—a major eruption could shoot lava more than half a mile into the air and create a suffocatingly hot ash plume. A disaster on that scale wouldn't be unheard of. In 79 CE, Italy's Mount Vesuvius erupted, covering the ancient Roman cities of Pompeii and Stabiae with hot ash and killing many of the inhabitants almost instantly.

A plaster cast of one of victims of the eruption of Vesuvius.

And in 1883 Krakatoa, a volcanic island in Indonesia, erupted with an explosion so loud that it was reported thousands of miles away. It's estimated that the size of that explosion was equivalent to that of thirteen thousand atomic bombs, and it killed more than thirty thousand people. In the aftermath of the eruption it was found that most of the island had disappeared.

These massive eruptions are not just part of our past; they are also, unfortunately, an inevitable part of our future. For instance, a massive buildup of lava in an undersea volcano off the south of Japan has recently been detected. The slow seepage of its lava has built up a dome two thousand feet above the seabed. The previous super-eruption in this volcanic area, seven thousand years ago, devastated the Japanese islands. Another such eruption may be brewing and is likely to have a similarly massive impact on Japan, as well as filling Earth's atmosphere with ash. This ash will end up in the atmosphere for years, blocking out the sun and lowering temperatures across the whole world, creating a so-called global winter.

But here's the strange thing. Despite billions of years of volcano eruptions and billions of years of tectonic movement, Earth's mountains aren't very high. The view of Earth from space shows this most strikingly; from up there, it seems we live on an almost perfect billiard ball, with nothing big poking out: the mountains are all relatively insignificant wrinkles on a smooth orb, yet they've had billions of years to grow tall — so why haven't they done so? Well, there are two processes that are constantly making the mountains smaller. The first is erosion: rain, ice, and winds are constantly rubbing small particles off mountains, weathering them and grinding them down. And in addition to this, while the weight of the mountains increases as they grow, it produces a pressure on the rock below that, over time, creeps and flows, driving the mountains back into the crust. So just as the ice sheets weigh down Antarctica, the mountains weigh down the tectonic plates from which they came, and the bigger they grow, the more they sink.

Of course the cabin crew didn't mention any of this as we were landing, which is perhaps the best way to deal with living on an unpredictable planet that's constantly on the move. For as much as we may understand the underlying causes of an earthquake, no one can predict when the next one will hit San Francisco. *Maybe it will be today,* I thought, looking across at Susan. She didn't seem worried. *She probably lives in denial,* I thought, *like the rest of us.* How else can we live happily on this thin crust, stretched out over a fluid planet that generates forces that are incomprehensibly big — so big that they've built mountains over millions of years, and destroyed whole cities in minutes; forces that have spewed new islands into existence and gobbled up others; forces that have caused whole continents to sink under the weight of ice, the very ice that's now melting, causing sea levels to rise inexorably and threatening all coastal cities, including San Francisco. And none of these forces are going to stop because they're all driven by the fluidity and liquidity of the planet. To survive as a civilization and as a species, we will have to learn how to live with them.

Susan was doing just that, using the camera on her phone to guide the application of her red lipstick. I liked her style. I still didn't know who she was, what made her tick, or where she was going. I only knew one thing for certain: her name really was Susan — I had read it off her customs form, which she'd filled in using my ballpoint pen. A pen that she had now taken with her as she nimbly edged out into the aisle, pulled her overhead baggage down in one fluid movement, and headed to the exit. Meanwhile, over the intercom we received one final optimistic message:

"On behalf of the airline and the entire crew, I'd like to thank you for joining us on this trip, and we look forward to seeing you on board again in the near future. Have a nice day!"

13 / SUSTAINABLE

LIVING ON A FLUID planet, the one thing we can be certain about is change: sea levels are rising; Earth's mantle is flowing, moving the continents; volcanoes erupt, creating new lands and destroying others; hurricanes, typhoons, and tsunamis continue to a batter our coastlines, reducing cities to rubble. In the face of this future it seems only rational to build our homes, roads, water systems, power stations, and indeed airports — all the stuff we rely on to live a dignified and civilized life — to withstand damage. This stuff needs to be strong and tough to survive earthquakes and floods, yes, but it'd be even better if we could design our infrastructure so it would repair itself, allowing our cities to be more nimble and resilient in the face of environmental change. This may sound far-fetched, but in fact, it's what biological systems have been doing for millions of years. Consider a tree: if it's damaged in a storm, it can repair itself by growing new limbs. Likewise, if you cut yourself, your skin heals itself. Could our cities become similarly self-healing?

In 1927 Professor Thomas Parnell of the University of Queensland in Australia conducted an experiment to see what would happen to black tar if it was left to settle in a funnel. What he found was this: Over days, it behaved like a solid, staying just where he put it. But over months and years it started to creep and behave like a liquid. Indeed, it flowed down the funnel and started to form droplets. The first drop fell in 1938, the second fell in 1947, the third in 1954, and so on, with the ninth drop falling in 2014. This is surprising behavior from a material that seems so solid when you drive over it in your car. That's asphalt, but asphalt is just tar mixed with stones. So what's going on?

The University of Queensland Pitch Drop experiment (taken in 1990, two years after the seventh drop, and ten years before the eighth drop fell).

Tar is a much more interesting material than anyone initially thought — materials scientists included. Extracted from the ground or produced as a byproduct of crude oil, it seems to be nothing more than boring black sludge. But in reality, it's a dynamic mixture of hydrocarbon molecules that formed over millions of years from the decayed molecular machinery of biological organisms. The decay products are complex molecules, which, although not part of a living system anymore, nevertheless self-organize within the tar, creating a set of interlinked structures. At normal temperatures the smaller molecules inside the tar have enough energy to move through its internal architecture, which gives the material fluidity. So tar is a liquid, albeit a very viscous one: it is two billion times more viscous than peanut butter, which explains why Professor Parnell's tar has taken so long to drip through the funnel.

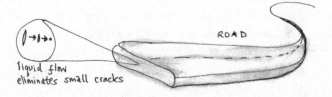

How the liquid flow inside asphalt roads allows cracks to heal themselves.

Tar's characteristic pungent smell comes from molecules that contain sulfur, an element often associated with smelly organic substances. When you walk or drive past engineers laying down a new road surface, you'll see and smell them heating up tar, which gives the molecules more energy to move, and thus to flow. But the extra energy also allows more of the molecules to evaporate into the air, and so the material becomes smellier, just as drinks become more aromatic when they are heated up.

A smelly liquid might seem an idiotic thing to build a road with, but engineers add stones to the material, creating a composite substance, part liquid and part solid — similar, in fact, to the structure of peanut butter, which is made of a lot of ground-up pieces of peanut all held together by an oil. The strength and hardness of the stones support the weight of vehicles driving over the asphalt and also helps the road resist damage from exposure. Cracks do sometimes open up if the forces exerted on the road get to be too high, but they do so between the stones and the tar that's gluing them together. This is where the liquid nature of tar comes to the rescue: the tar flows in and reseals these cracks, allowing the road to repair itself and last far longer than a purely solid surface ever would.

Of course, as road users, you'll have noticed that there is a limit to their self-repairing properties: roads eventually do get old and start to disintegrate. Temperature is partially to blame. If the temperature gets below, say, 68°F, then the liquid tar becomes so viscous that it cannot reflow and heal the cracks as they appear. And beyond that, over time, oxygen from the air reacts with molecules on the surface of the tar and alters their properties, again making it more and more viscous, and less and less able to seal up cracks. Over time, the road skin will change color and become less fluid, just as your skin becomes less flexible and drier as you age. This is when you'll see small potholes form, which, unless tended to, grow and grow and eventually destroy the road surface entirely.

A case in point: my journey on the airport shuttle bus to the hotel. As soon as we arrived in the city, we got stuck in a traffic jam caused by lane closures due to road resurfacing. The shuttle crawled along as three lanes converged into one — by my estimation, we moved less than a mile in thirty minutes. It was 2 a.m. according to my body clock; I was tired and I desperately needed to pee.

It doesn't have to be this way. Or at least that's what we materials scientists hope. Scientists and engineers around the world are busily developing strategies for increasing the life of roads, and thus reducing traffic jams. In the Netherlands, a group of engineers is studying the effect of incorporating tiny microscopic fibers of steel into tar. This does not alter the mechanical properties of the road much, but it does make them more powerful. When the material is exposed to an alternating magnetic field, electric currents flow inside the steel fibers, heating them up. The hot steel, in turn, heats up the tar, making it locally more fluid, allowing it to flow and heal any cracks. Essentially, they're supercharging the self-healing properties of tar and also countering the challenges of winter's cold temperatures. The technology is now being tested on stretches of motorway in the Netherlands, using a special vehicle that applies a magnetic field to the road as it drives along. The idea is that, in the future, all vehicles could be fitted with such a device, so anyone driving on a road would also be revitalizing it.

Another way to address tar's natural loss of fluidity is to replenish its lost volatile ingredients — the molecules that make it flow. The easiest way of doing that would be to apply a special kind of cream to the road surface — essentially a moisturizing cream, just like the ones we use on our skin. A more sophisticated version of this method is being tested by a group in Nottingham University in the UK, led by Dr. Alvaro García. They put microcapsules of sunflower oil into tar. These remain intact inside the material until microcracks form, causing the capsules to rupture. The oil, once released, locally increases the fluidity of the tar and thus promotes flow and self-healing capabilities. The results of these studies show that cracked asphalt samples are restored to their full strength two days after the sunflower oil is released. This is a dramatic improvement. It is estimated that this has the potential to increase the lifespan of a road from twelve years to sixteen years with only a marginal increase in cost.

In our research group at the Institute of Making, we are working on technologies that can help repair asphalt efficiently once the cracks have already got bigger: we've started to do 3D printing of tar.

3D printing is a relatively new way of making and repairing objects. Thousands of years ago, printing was invented in China as a process of transferring ink onto a page via a wooden printing block. The rest of the world caught on and innovated, giving us a world of books, newspapers, and magazines — an information

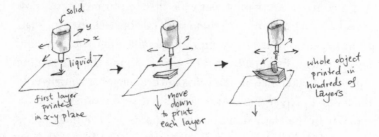

The 3D printing process. A print head converts a solid to a liquid (often by heating) and squirts it out in a predetermined pattern in an x-y plane. Once cooled, this creates a single solid layer. Then the printing platform is moved down and another layer in a different pattern is printed. Printing hundreds of layers this way creates a whole object.

revolution. But all that is 2D printing; 3D printing takes the approach one stage further. Instead of printing a thin, two-dimensional layer of liquid ink onto a page, 3D printing allows you to print many two-dimensional layers of liquid on top of one another, each one solidifying before the next is applied, ultimately constructing a 3D object.

Of course, you don't need to use ink to make a 3D print. You can use any material that can be transformed from a liquid into a solid. Just look at bees. This is exactly how they make their extraordinary hexagonal honeycombs. When they are between twelve and twenty days old, worker bees develop a special gland for converting honey into soft wax flakes. They chew up the wax and deposit it layer by layer to make the honeycomb. Wasps use the same trick to make their nests, chewing wood fibers and mixing them with saliva to create paper houses for their larvae.

Human 3D printing technology is now catching up with the bees and wasps. Plastics, for instance, can be squirted out of a printer, layer by layer, to create solid objects more complex than honeycombs. Even objects with moving parts can be printed—this technique is being used in medicine to create prosthetics with

Bees were using 3D printing to build their honeycombs long before humans happened upon the technique.

functioning joints, all made in one piece, at low cost. Also, 3D printing can be used to print biological materials. In 2018 Chinese scientists conducted the first clinical trials to create replacement ears for children suffering from birth deformities. They did this using the children's own cell tissue and 3D printers to create the scaffold for the cells to grow into ears.

3D printing works for metals too. The Dutch company MX3D is using 3D printing to make steel bridges, adding molten steel, blob by blob, and relying on techniques borrowed from welding technology. Another technique for the 3D printing of metal objects is to use a high-powered laser that melts metal powders and joins them. This process is being used to make everything from gold jewelry to jet engine parts. One of the major advantages of this technique is that it's easy to make things hollow, which lowers weight and saves material. Objects are increasingly being designed to have arteries, allowing coolant, lubricant, or even fuel to flow through them. In essence, this design mimics our bodies — we are part solid flesh and part liquid. Our blood delivers nutrients via our circulatory and arterial systems, which also deliver proteins and other molecular ingredients to parts of our body that are hurt, allowing them to grow new cells to replace damaged ones in our skin, brain, liver, kidneys, heart, and so on. This is another aspect of nature we can now emulate, thanks to 3D printing, potentially allowing technology to last longer by repairing itself, and so be more sustainable.

The byproduct, of course, of the body's reliance on circulating fluids is that it creates waste, which also needs to be ejected. Getting rid of some liquid was the first thing on my mind when I got off the shuttle bus in front of my hotel in San Francisco: I still needed to pee badly. I hopped from foot to foot during check-in, and then sprinted to my allocated room, almost wetting myself as I spent frustrating moments swiping, and re-swiping, my pass card in the slot of the door, until I finally got it open. Then, oh, the relief!

. The delight of an en suite bathroom goes far beyond being able to pee at will. It is the place we go to clean, to be refreshed, and to luxuriate. And it all depends on the availability of free-flowing clean water. Most people in developed counties take this for granted because almost none of the infrastructure that delivers our water and removes our waste is visible. But it is there, a vital network within our cities and surprisingly expensive to run, even in places like San Francisco, where water is so plentiful. Keeping waste contained and cleaning it up so it can be returned to our rivers and seas without polluting them require a lot of filtering machines, settling tanks, and reprocessing units. All of which costs money and energy. The less effluent you want out there polluting ecosystems, the more the effort costs, and the more water you need to dilute whatever's coming out of reprocessing plants. So dealing with the wastewater from dishwashers, washing machines, showers, baths, and toilets for a city the size of San Francisco is not easy. The drinking water supply also has to come from somewhere, which requires more filtering, pumping, and monitoring. Every time water goes round the loop, from clean to dirty and back again, it costs energy and creates an environmental impact through the creation of waste products.

Manufacturing also uses huge amounts of water, and so, in buying most products, you're also increasing your so-called water footprint. You may be someone who has a shower only twice a week and uses a low-flush toilet, but your water footprint is still likely to be substantial. It is estimated that the average American's water footprint for goods they purchase, and use only once, is 583 gallons per day, thanks to water-intensive goods like paper, meat, and textiles. Even seemingly mundane activities such as eating a hamburger, reading the newspaper, and buying a T-shirt have a big impact on a person's water footprint. Hence the sign in the hotel bathroom, reminding me that water is a valuable resource and encouraging me not to request new towels every day.

As the world population increases to ten billion over the next few decades, it is estimated that access to clean water will be an increasing struggle in many parts of the world. Currently one billion people lack access to clean water, and a third of the global population experiences shortages through the year. Without access to clean water, we can expect an increase in poverty, malnutrition, and the spread of disease. It should be emphasized that this issue affects big cities as well as rural communities. For instance, the Brazilian city of São Paolo experienced severe water shortages in 2015 when drought emptied its main reservoir. At the worst point of the crisis, it was estimated that the city, with a population of 21.7 million people, had only twenty days of water left. Many other megacities around the globe face similar problems imposed by variations in climate, growth in population, and, as wealth increases, a bigger water footprint per person.

While we obviously all rely on water, a sustainable and healthy society needs other liquids too. Some are surprising. For instance, liquid glass. A lot of our food and drink is preserved and transported in glass bottles or jars. It's a great material for that. Chemically inert, glass doesn't react with the contents it holds, so those products last longer. But glass does break, and when it does, it has to be melted down into a liquid again in order to be remade into another vessel. This has been happening for thousands of years: a circular system that allows us to reuse waste.

Glass, as a material for food and drink containers, does have its downsides. It is dense, so transporting it around the world costs a lot of energy. Also, remelting it requires a lot of energy because glass has such a high melting point. Because of these two factors, in a world powered largely by fossil fuels, glass containers end up exacerbating the problems caused by climate change.

Hence the shift in the twentieth century toward plastic packaging, which is lighter and more flexible and requires a great deal less energy to remelt into new packaging. That's the theory, anyway. The reality is quite different. Many, many different packaging plastics have been developed, each one amazing in its ability to preserve and package foods, liquids, electronics, and more. What no one thought through was what would happen if these plastics were all collected, recycled, and melted together. The mixture creates inferior plastic, unable to perform the jobs of the originals because the individual hydrocarbon molecules that make up a typical plastic chemically bond to one another in particular ways. This bonding creates certain structures inside the plastic, which determine its strength, elasticity, and transparency. If you melt together different plastics, you end up with a mess. Thus the plastics need to be carefully uncooked to make them usable again. Since there are more than two hundred plastics in common use, and any number of items on the market are packaged with two or three different types, all in a rainbow of colors, separating plastics has become a costly task. We haven't yet found a way of liquefying them to create a sustainable system.

Sadly, worldwide, the majority of plastic packaging is not recycled, a fact that is steadily creating an environmental disaster. Our landfills are overflowing with plastics, and because plastic packaging is designed to be lightweight, it's easily carried off by the wind. And because plastics float, when they land in a river, they eventually make their way to the seas and oceans, contaminating those ecosystems. This is happening at an ever-increasing pace. At current rates, it is estimated that by 2050 there will be more plastic than fish in the oceans.

There is no easy answer to the problem of plastic packaging. Using glass, as already mentioned, requires a lot energy, which, unless it is generated with renewable power sources, is unsustainable. Paper is another possible replacement, but its production is more energy-and-water-intensive than plastic's. Using less packaging is an attractive possibility. But since most agriculture and manufacturing is highly water-intensive, less packaging may result in greater waste. Overall, this could easily put more pressure on global water and food supplies. Thus we find the problem of sustainable packaging has flowed full circle, as is often the way with things that rely on liquids.

And so I was expecting a lot from this conference on sustainable technology, which I had flown five thousand miles to participate in. Would the attendees be interested in our work on self-repairing cities and the 3D printing of tar, or would the discussion focus on cheaper ways to desalinate water or create sustainable packaging? Either way, I knew that understanding the behavior of liquids was going to be essential. I looked at my watch. The opening talk of the conference would be starting soon. I splashed some water on my face to stave off jet lag and headed downstairs to the convention center.

When I got there, I saw something I had not expected: Susan, striding onto the stage. My eyes almost popped out of my head. This person whom I knew so well — having spent eleven hours sitting next to her — was an engineer. And not just any engineer, but a keynote speaker of the very meeting I had flown halfway around the world to attend. She talked brilliantly, and wide-rangingly, about the complex global sustainability challenges facing us. But in truth I found it hard to concentrate, as I was so furious with myself for not talking to her on the plane.

After Susan's presentation, I couldn't resist going up to talk to her. I had to wait in line as she patiently dealt with the others crowding around her. When my turn came, I smiled and, trying to sound cool, said, "Nice talk."

She looked at me, puzzled for a second, apparently trying to recall how she knew me. But then it clicked. "I suppose you want your pen back," she said.

EPILOGUE

As I hope the account of my journey from London to San Francisco has shown, an airplane flight is made possible, and delightful, by our understanding and control of myriad liquids, from kerosene to coffee, from epoxies to liquid crystals. There are many liquids I haven't mentioned, but I wasn't trying to be comprehensive. Instead, I've tried to paint a picture of our relationship with liquids. For thousands of years we have been trying to come to grips with this alluring yet sinister, refreshing yet slimy, life-giving yet explosive, delicious yet poisonous state of matter. So far we have largely managed to harness the power of liquids while protecting ourselves from their dangers (tsunamis and rising sea levels notwithstanding). Looking ahead, I'm guessing that our future will be just as liquid-filled as our past, but our relationship with liquids will deepen.

Take medicine, for instance. Most medical tests require a sample of blood or saliva, which doctors use to diagnose illness or monitor health. These tests almost always have to be done in a lab, and they're both time-consuming and costly. They also require a visit to a doctor or a hospital, which is not always possible, especially in countries where medical resources are scarce. But a new technology called lab-on-a-chip is likely to change all this, ushering in a future in which diagnostics are carried out at home, almost instantly, and cheaply.

Lab-on-a-chip technology allows you to take small samples of your own bodily fluids and feed them into a small machine that examines their biochemical composition. These chips process liquids in much the same way that silicon microchips process digital information. Your blood, or whichever other fluid, is directed into a series of microscopic internal tubes, which can divert droplets in different directions, to undergo different kinds of analysis. It's still early days for these chips, but be prepared to hear more and more about them in the coming years. With the potential to diagnose everything from heart disease to bacterial infection to early-stage cancer, they're likely to be at the foreground of a revolution in medical technology akin to what we've seen in the IT industry — but this time, the revolution will be liquid.

In order for lab-on-a-chip technology to work, it has to have a mechanism that allows it to move and manipulate small droplets of liquid. Biological organisms, of course, are expert at this. Go into a garden during a rain shower and you'll see leaves that repel water so effectively that the raindrops bounce off. Lotus leaves, for instance, have long been known to have this super-hydrophobic property, but no one knew why until quite recently, when electron microscopes revealed something odd about their surface. As suspected, they're coated in a waxy material that repels water, but surprisingly, that material is arranged on the surface in the form of billions of tiny microscopic bumps. When a drop of water sits on this waxy surface, it tries to minimize its area of contact, because of the high surface tension between drop and surface. The bumps on the lotus leaf drastically increase this area of waxiness, forcing the droplet to sit up precariously on the tips of the bumps. In this state, the droplet becomes mobile and quickly slides off the leaf, along the way collecting small particles of dust, hoovering them up like a mini–vacuum cleaner: a lotus leaf's secret to staying shiny and clean.

Manipulating material surfaces to make them super-hydrophobic is likely to become big business in the coming years. This will allow us not only to guide droplets through the internal workings of lab-on-a-chip technology, but also to do a great many other things too. We will, for instance, be able to keep water from sticking to windows, thus keeping them as clean as a lotus leaf. We might also be able to develop waterproof clothing that harvests the water that falls on it, transporting it through tiny tubes into a collecting pouch, so it can be drunk later. This design is inspired by the thorny devil lizard, which hydrates itself by collecting any rainwater that falls on its skin, manipulating it through tiny channels by means of capillary flow.

The thorny devil lizard collects water through its skin by using hydrophobic materials and capillary flow.

The potential of such water-collection technology for the billions of people without access to regular supplies of clean water is huge, especially if a cheap way to filter the water can also be mastered. A new material that may be capable of this is called graphene oxide. It is a two-dimensional layer of carbon and oxygen atoms. In the form of a membrane it acts as a barrier layer for most types of chemical molecules but easily allows water molecules through. So it is like a molecular sieve. Potentially it could make an extremely effective and cheap water filter, which could even make seawater drinkable.

As we know, water is a life-giving substance, and it's generally accepted that the presence of liquid water is what allowed life on Earth to evolve from very basic chemical structures to the complex cells we are made of. But this is still a hypothesis; we don't really know with any certainty how it happened. Scientists all over the world are doing experiments in an attempt to work it out by re-creating the chemical conditions that were present when life evolved here four billion years ago. It seems most likely, at this point, that life originated at the bottom of our deep oceans. There, thermal vents create a complex chemical soup with many of the ingredients that we find in our cells. As the twenty-first century progresses, exploring these regions, and the deep sea in general, will be an important frontier for us. It's odd, really, that we know less about the bottom of our own oceans than we do about the surface of the moon.

If the depths of the ocean are our next physical frontier, I'd say that we have two computational ones on the horizon, both dependent on liquid. Cells and computers both compute information, but in completely different ways. Cells function and reproduce by computing the information stored in DNA via chemical reactions. Silicon-based computers, by contrast, read chips that contain billions of solid transistors, which react to incoming electrical signals transcribed from a computer program. The signals are communicated through a series of 1s and 0s, the binary language of digital computers. The transistors apply logic to the flow of 1s and 0s, computing answers again in the form of 1s and 0s and moving them to another part of the computer chip. It all may seem very basic, but by doing billions of simple calculations a second, sophisticated computation can be performed—the sort that beats chess grandmasters and computes the trajectory of a rocket to the moon.

When cells compute things, they use chemical reactions instead of transistors. Instead of 1s and 0s, they compute with molecules, and they communicate with molecules. There are no transistors or wires, just chemical reactions in the liquid state, swimming around inside the cells. These chemical reactions happen incredibly fast, and simultaneously, all over the cell, making this so-called parallel computing system extremely efficient. The molecules involved are also all very small — you can easily have a sextillion (1,000,000,000,000,000,000,000) molecules in a single drop of liquid, a potentially colossal source of computational power and memory.

Scientists are trying to emulate this process by using DNA to create a liquid computer. The work is developing rapidly, especially as ways to manipulate DNA and do calculations in test tubes are becoming more and more sophisticated and readily available. In 2013 researchers hit a big milestone: they were able to store a digital photograph's data in liquid and then retrieve it. This opens the door to a whole new paradigm of computation — in the future, you might be able to store all your data in a single drop of liquid.

Liquid computing is the first of the incredible computational systems that are in development. The second is quantum computing, which relies on the quantum versions of binary's 1s and 0s — meaning a piece of information is stored in the computer as both a 1 and a 0 until a computation has been completed. Quantum computing takes advantage of the rules of quantum mechanics, which allow all possible outcomes of an event to exist simultaneously. As such, all possible answers to a problem can be computed at once, vastly speeding up calculations. There are machines already that can do this, but they're still quite rudimentary. One thing is certain, though: to run, they rely on very cold temperatures that can be obtained only with the help of a very special liquid: liquid helium.

Helium is a gas until cooled to −452.2°F; at this temperature, which is just seven degrees above absolute zero, it turns into a liquid. Luckily, we already have a sense of how to work with liquid helium, thanks to hospital equipment. If you've ever had a brain, hip, knee, or ankle injury, or been diagnosed with cancer, you've most likely had an MRI scan. But without super-cold liquid helium, these diagnostic tools, vital to all modern hospitals, would simply cease to operate. The cold temperatures of liquid helium are what make it possible for MRI machines to reliably detect tiny changes in the magnetic fields inside the human body and so map our internal organs. Unfortunately, though, although helium is one of the most abundant elements in the universe, it is quite rare on Earth. Hospital shortages of liquid helium now occur quite regularly and supplies often run out. In response, geologists are constantly prospecting for new sources of helium in Earth's crust (usually found in natural gas), but because of its growing importance, prices for this crucial substance have risen 500 percent in the past fifteen years.

As useful as liquid helium is, it's also quite unruly. It will operate successfully to cool MRI machines to −452°F, but cool it down a few more degrees to −457°F, and it enters what we call a superfluid state. In this state, all the atoms in the liquid occupy a single quantum state, which is to say that all of the billions and billions of helium molecules act as if they are a single molecule, giving the liquid odd powers — it has no viscosity, for instance, and will spontaneously flow up out of a container. It will even be able to flow through solid materials, finding its way through the object's atomic-sized defects without any resistance.

By this point in the book, I hope, this sort of behavior won't be so surprising to you. Liquids have a duality: they are neither a gas nor a solid, but something in between. They are exciting and powerful, on the one hand, while anarchic and slightly terrifying on the other. That is their nature. Nevertheless, our ability to control liquids has mostly yielded a positive impact for humanity, and my bet is that at the end of the twenty-first century we'll look back at lab-on-a-chip medical diagnostics and cheap water desalination and hail them as major breakthroughs that made longer life expectancies possible and prevented mass migrations and conflict. By then I also hope we'll have said goodbye to burning fossil fuels — in particular, kerosene. This liquid has given us the gift of cheap global travel, of sunny holidays and exciting adventures, but its role in global warming is too great to ignore. What liquid will we invent to replace it? Whatever it is, I suspect we will never dispense with the preflight safety ritual. Perhaps it will no longer involve the props of life jackets, oxygen masks, and seat belts — but we will always need ceremonies to celebrate the dangerous and delightful power of liquids.

ACKNOWLEDGMENTS

I sincerely thank my editors, Daniel Crewe and Naomi Gibbs, for being so patient, supportive, and critically incisive, and for putting up with my obsession with the preflight safety briefing.

I work at the Institute of Making with a team of scientists, artists, makers, engineers, archaeologists, designers, and anthropologists. They have all helped me in some way to make this book. I want to thank the whole team for their friendship and support: Zoe Laughlin, Martin Conreen, Ellie Doney, Sarah Wilkes, George Walker, Darren Ellis, Romain Meunier, Necole Schmitz, Elizabeth Corbin, Sara Brouwer, Beth Munro, and Anna Ploszjaski.

The Institute of Making is part of UCL, a university that nurtures teaching and research across disciplines. There are many colleagues who make it the intellectually vibrant place it is, and I want to thank in particular Buzz Baum, Andrea Sella, Guillaume Charras, Yiannis Ventikos, Mykal Riley, Mark Lythgoe, Helen Czerski, Rebecca Shipley, David Price, Nick Tyler, Matthew Beaumont, Nigel Titchener-Hooker, Marc-Olivier Coppens, Paola Lettieri, Anthony Finkelstein, Polina Bayvel, Cathy Holloway, Richard Catlow, Nick Lane, Aarathi Prasad, Manish Tiware, Richard Jackson, Mark Ransley, and Ben Oldfrey.

The UK has a particularly vibrant science and engineering community, which it has been a pleasure to be part of for so many years. I am grateful for support particularly from Mike Ashby, Athene Donald, Molly Stevens, Peter Haynes, Adrian Sutton, Chris Lorenz, Jess Wade, Jason Reese, Raul Fuentes, Phil Purnell, Rob

Richardson, Iain Todd, Brian Derby, Marcus Du Sautoy, Jim Al Khalili, Alom Shaha, Alok Jha, Olivia Clemence, Olympia Brown, Gail Cadrew, Suze Kundu, Andres Tretiakov, Alice Roberts, Greg Foot, Timandra Harkness, Gina Collins, Roger Highfield, Vivienne Parry, Hannah Devlin, and Rhys Morgan.

I would particularly like to thank those who commented on the book as it took shape: Ian Hamilton, Sally Day, John Comisi, Rhys Phillips, Clare Pettit and Sarah Wilkes. Andrea Sella, Philip Ball, Sophie Miodownik, Aron Miodownik, Buzz Baum, and Enrico Coen all read full drafts of the book and gave me extremely helpful feedback.

I'd like to thank my literary agent, Peter Tallack, who got the book off the ground in the first place, and the teams at Penguin Random House and Houghton Mifflin Harcourt for help with the production process.

I am very grateful to Lal Hitchcock, George Wright, and Diane Storey, for all the support, and the many wonderful Dorset days together while writing this book.

I'd like to thank my kids, Lazlo and Ida, for sharing their boundless enthusiasm for liquids and helping me with the very entertaining experimental phase of this book.

And, finally, I'd like to thank my love Ruby Wright, for being my editor in chief and my creative inspiration.

FURTHER READING

Ball, Philip, *Bright Earth: Art and the Invention of Colour*, Vintage Books (2001).

Faraday, Michael, *The Chemical History of a Candle*, Oxford University Press (2011).

Jha, Alok, *The Water Book*, Headline (2016).

Melville, Herman, *Moby-Dick*, Penguin Books (2001).

Mitov, Michel, *Sensitive Matter: Foams, Gels, Liquid Crystals, and Other Miracles*, Harvard University Press (2012).

Pretor-Pinney, Gavin, *The Cloudspotter's Guide*, Sceptre (2007).

Roach, Mary, *Gulp: Adventures on the Alimentary Canal*, Oneworld (2013).

Rogers, Adam, *Proof: The Science of Booze*, Mariner Books (2015).

Salsburg, David, *The Lady Tasting Tea: How Statistics Revolutionized Science in the Twentieth Century*, Holt McDougal (2012).

Spence, Charles, and Bentina Piqueras-Fiszman, *The Perfect Meal: The Multisensory Science of Food and Dining*, Wiley-Blackwell (2014).

Standage, Tom, *A History of the World in Six Glasses*, Walker (2005).

Vanhoenacker, Mark, *Skyfaring: A Journey with a Pilot*, Chatto & Windus (2015).

PICTURE CREDITS

INDEX

KI

c.1